SCIENTISTS OF WALES

William Robert Grove

SCIENTISTS OF WALES

Series Editor
Professor Gareth Ffowc Roberts
Bangor University

Editorial Panel
John V. Tucker
Swansea University

Iwan Rhys Morus
Aberystwyth University

SCIENTISTS OF WALES

William Robert Grove

VICTORIAN GENTLEMAN OF SCIENCE

IWAN RHYS MORUS

UNIVERSITY OF WALES PRESS
2017

© Iwan Rhys Morus, 2017

All rights reserved. No part of this book may be reproduced in any material form (including photocopying or storing it in any medium by electronic means and whether or not transiently or incidentally to some other use of this publication) without the written permission of the copyright owner except in accordance with the provisions of the Copyright, Designs and Patents Act 1988. Applications for the copyright owner's written permission to reproduce any part of this publication should be addressed to the University of Wales Press, 10 Columbus Walk, Brigantine Place, Cardiff CF10 4UP.

www.uwp.co.uk

British Library Cataloguing-in-Publication Data
A catalogue record for this book is available from the British Library.

ISBN   978-1-78683-020-3 hardback
        978-1-78683-004-3 paperback
eISBN   978-1-78683-005-0

The right of Iwan Rhys Morus to be identified as author of this work has been asserted in accordance with sections 77, 78 and 79 of the Copyright, Designs and Patents Act 1988.

Typeset by Marie Doherty
Printed by CPI Antony Rowe, Melksham.

i fy nhad

CONTENTS

Series Editor's Foreword   ix
List of Illustrations   xi

Prologue   1

1. A Scientific People   7
2. The Metropolis of Science   23
3. The Correlation of Physical Forces   43
4. Scientific Reform   65
5. Swansea Science   85
6. Unifying Science   103
7. A Scientific Statesman   121

Afterword   139
Notes   147
Bibliography   161
Index   169

SERIES EDITOR'S FOREWORD

Wales has a long and important history of contributions to scientific and technological discovery and innovation stretching from the Middle Ages to the present day. From medieval scholars to contemporary scientists and engineers, Welsh individuals have been at the forefront of efforts to understand and control the world around us. For much of Welsh history, science has played a key role in Welsh culture: bards drew on scientific ideas in their poetry; renaissance gentlemen devoted themselves to natural history; the leaders of early Welsh Methodism filled their hymns with scientific references. During the nineteenth century, scientific societies flourished and Wales was transformed by engineering and technology. In the twentieth century the work of Welsh scientists continued to influence developments in their fields.

Much of this exciting and vibrant Welsh scientific history has now disappeared from historical memory. The aim of the Scientists of Wales series is to resurrect the role of science and technology in Welsh history. Its volumes trace the careers and achievements of Welsh investigators, setting their work within their cultural contexts. They demonstrate how scientists and engineers have contributed to the making of modern Wales as well as showing the ways in which Wales has played a crucial role in the emergence of modern science and engineering.

RHAGAIR GOLYGYDD Y GYFRES

O'r Oesoedd Canol hyd heddiw, mae gan Gymru hanes hir a phwysig o gyfrannu at ddarganfyddiadau a menter gwyddonol a thechnolegol. O'r ysgolheigion cynharaf i wyddonwyr a pheirianwyr cyfoes, mae Cymry wedi bod yn flaenllaw yn yr ymdrech i ddeall a rheoli'r byd o'n cwmpas. Mae gwyddoniaeth wedi chwarae rôl allweddol o fewn diwylliant Cymreig am ran helaeth o hanes Cymru: arferai'r beirdd llys dynnu ar syniadau gwyddonol yn eu barddoniaeth; roedd gan wŷr y Dadeni ddiddordeb brwd yn y gwyddorau naturiol; ac roedd emynau arweinwyr cynnar Methodistiaeth Gymreig yn llawn cyfeiriadau gwyddonol. Blodeuodd cymdeithasau gwyddonol yn ystod y bedwaredd ganrif ar bymtheg, a thrawsffurfiwyd Cymru gan beirianneg a thechnoleg. Ac, yn ogystal, bu gwyddonwyr Cymreig yn ddylanwadol mewn sawl maes gwyddonol a thechnolegol yn yr ugeinfed ganrif.

Mae llawer o'r hanes gwyddonol Cymreig cyffrous yma wedi hen ddiflannu. Amcan cyfres Gwyddonwyr Cymru yw i danlinellu cyfraniad gwyddoniaeth a thechnoleg yn hanes Cymru, â'i chyfrolau'n olrhain gyrfaoedd a champau gwyddonwyr Cymreig gan osod eu gwaith yn ei gyd-destun diwylliannol. Trwy ddangos sut y cyfrannodd gwyddonwyr a pheirianwyr at greu'r Gymru fodern, dadlennir hefyd sut y mae Cymru wedi chwarae rhan hanfodol yn natblygiad gwyddoniaeth a pheirianneg fodern.

LIST OF ILLUSTRATIONS

Figure 1   William Robert Grove, *Portraits of Men of Eminence*   41
           in Literature, Science, and Art, with Biographical Memoirs,
           the Photographs from Life, by Ernest Edwards, BA
           (London: Alfred William Bennett, 1865).
Figure 2   The Gas Battery, *Philosophical Transactions*, 1843.   55
Figure 3   Somerset House, Meeting of the Royal Society,   73
           engraved by H. Melville after a picture by Fairholt,
           published in *London Interiors*, 1845.
Figure 4   Map of locations for the Swansea meeting of the   93
           British Association for the Advancement of Science,
           1848.
Figure 5   A Soirée at the Swansea meeting of the British   99
           Association for the Advancement of Science, drawing
           by John Weir Padley. National Library of Wales.
Figure 6   Discharge experiments, *Philosophical Transactions*,   117
           1858.
Figure 7   Cartoon of Mr Justice Grove, from *Vanity Fair*,   135
           1887.
Figure 8   Frontispiece of Henry M. Noad, *Lectures on*   142
           Electricity, 1844.

PROLOGUE

It is quite easy to imagine an alternative steampunk universe in which William Robert Grove is celebrated as the inaugurator of a second Victorian industrial revolution. In this parallel world, his invention of the gas battery in 1841 revolutionised transportation, and killed off the steam locomotive before it had a chance to monopolise the railways. Banks of gas batteries, fuelled by cheap hydrogen produced from coal, powered horseless carriages on the roads as well as the railways, making possible fast and reliable travel beyond the confines of the rail network. The cheap power generated from coal-produced hydrogen quickly made electricity a serious competitor for steam in a variety of industries too. Crucially, the gas battery brought electricity into the Victorian home as well. By the 1860s, large gasometers filled with hydrogen were a common sight in Victorian towns and cities. Networks of pipes carried the gas to middle-class homes to provide fuel for gas batteries that powered not only electric lights but the whole range of electrical gadgetry that packed Victorian houses. Appliances for washing dishes, doing the laundry, cleaning carpets and curtains, even cooking, were all powered using Grove's ubiquitous gas batteries. They operated the magic lanterns that provided middle-class Victorians with their evening's domestic entertainment.

None of this happened, of course. But let us imagine another alternative history instead, observed this time from the perspective of the twenty-second century. In this history, Grove is the hero again. During the early decades of the twenty-first century – nearly two centuries after it was invented – Grove's gas battery, now called the fuel cell, transformed transportation. It provided a solution to increasingly urgent problems of climate change and the pollution produced by hydrocarbon-based fossil fuels. Cars and lorries powered by fuel cells, running on hydrogen gas generated by electrolysis using solar power, inaugurated a revolution in low-emission travel. The only by-product of combining hydrogen and oxygen to make electricity is water, after all. The fuel cell generated none of the greenhouse gases that bedevilled the traditional petrol- or diesel-powered internal combustion engine. It helped too that it ran on fuel that could be found anywhere there was water, rather than one derived from increasingly scarce and expensive to acquire oil. Grove's gas battery therefore played a key role in the battle to reduce environmentally harmful emissions and contain climate change that dominated the second half of the twenty-first century and Grove himself became an icon for the drive towards clean energy.

This has not happened yet, of course. But there are signs already that this really is one plausible route that the history of twenty-first-century energy use might follow. In the years immediately following his invention, Grove's gas battery received very little attention other than as a philosophical curiosity. It certainly did not strike anyone at the time – Grove included – as a potential source of power on an industrial scale, even though many electricians at the time already regarded electric power as a viable competitor for steam. Grove did suggest in 1854,

> that if, instead of using zinc and acids, which are manufactured, and comparatively expensive materials, for the production of electricity, we could realize the electricity developed by the combustion in atmospheric air, of common coal, wood, fat, or other raw material, we should have at once a fair prospect of the commercial application of electricity.[1]

He did nothing about it though. Others such as Lord Rayleigh and Ludwig Mond did make attempts to turn the gas battery into a practical, commercially viable technology but again they went nowhere.[2] Now, on the other hand, car manufacturers such as Toyota are pouring significant resources into developing fuel cell-powered electric vehicles. Grove's gas battery seems set to become the technology of the future.

I am putting these two sketches of alternative and future histories forward here, however, because thinking about them might help us understand the current place in history of this biography's subject, the Welsh Victorian man of science William Robert Grove. Grove is in many ways a curious historical figure. For much of his own lifetime he was widely considered by his contemporaries to be one of the Victorian Age's leading men of science. By the end of his life, however, he was an increasingly marginal figure as far as science was concerned – mainly recognized for his later career as barrister and judge. Since then he has been largely forgotten, except by the small number of historians of Victorian science. His life and scientific work are now starting to attract renewed attention, both from historians and the wider public. The reason for this is straightforward: the fuel cell is now in the headlines as a possible alternative source of energy for twenty-first-century transport. If pundits' prognostications come true, Grove's reputation and history a century from now may indeed be very different. And what would the history of science – and of Wales – look like if a Welshman were credited with inventing one this century's most significant technological innovations? What would that do to the way we – the Welsh – think about ourselves and science?

William Robert Grove was a key figure in the world of Victorian science. His early reputation as a man of science was indeed based on his invention of a battery too – though not the gas battery, as it happens. Grove hit the early Victorian headlines as the inventor of the nitric acid battery, notable for its constant flow of electricity and its power. It was soon to be a vital piece of equipment for the burgeoning telegraph industry. His essay *On the Correlation of Physical Forces*, published in 1846 was widely read and went through six editions over the following few decades. Most importantly maybe, he was known as a scientific statesman who played a key role in the reform of scientific institutions during the 1840s and 1850s. Grove was well known enough for his name to be casually dropped in improving Victorian children's books.

> You do not know what Nature is, or what she can do; and nobody knows; not even Sir Roderick Murchison, or Professor Huxley, or Mr. Darwin, or Professor Faraday, or Mr. Grove, or any other of the great men whom good boys are taught to respect

Or so said Charles Kingsley in *The Water Babies*.[3] In view of the way he was regarded by his contemporaries Grove is surprisingly understudied. He was accorded a chapter in J. G. Crowther's *Statesmen of Science*, but there are no other substantial biographies.[4]

In fact Grove was one of a handful of Victorian men of science who were famous then and almost forgotten now. John Tyndall or John Herschel might be other examples. As with Herschel, one reason, at least, why Grove went out of fashion was that his scientific ideas did too. By the end of his life, the correlation of physical forces had been replaced by the conservation of energy. The tide of physics had overtaken him. The example of Peter Guthrie Tait, co-author of one of the leading physics textbooks of the second half of the century, is instructive in this respect. Tait had clearly read Grove's book before delivering his inaugural lecture as professor of natural philosophy at Edinburgh in 1860. He offered his audience a careful account of Grove's theory regarding the mutual relations of the natural forces, and discussed 'what has been called the CORRELATION OF THE PHYSICAL FORCES'. Only two years later he was writing to his textbook co-author William Thomson asking if he had 'read Grove's Book' and whether he too 'feel all along the impression of humbug'.[5] Beyond its role as an energy source in space exploration, Grove's gas battery remained little more than a chemical curiosity for more than a century after its invention. Only in the last decade or so has it started to look like a realistic solution to pressing environmental problems.

Grove himself certainly did not regard his gas battery as a panacea for energy or environmental deficiencies. He recognised its potential utility, but just like other natural philosophers such as James Prescott Joule, he was dubious about the extent of electricity's usefulness more generally as a source of power. Inevitably, Grove thought, as long as 'with the one we use for fuel manufactured materials, in the production of which coals, labour, &c., have been expended; in the other, coals and water are used directly', electricity from a battery could never trump steam in the economic game.[6] But he also knew that there were other ways of doing it. If, for example,

> instead of employing manufactured products or educts, such as zinc and acids, we could realise as electricity the whole of the chemical force which is active in the combustion of cheap and abundant raw materials ... we should obtain one of the greatest practical desiderata, and have at our command a mechanical power in every respect superior in its applicability to the steam-engine.[7]

He told the British Association for the Advancement of Science (BAAS) in 1866 that 'we are at present, far from seeing a practical mode of replacing that granary of force, the coal-fields; but we may with confidence rely on invention being in this case, as in others, born of necessity, when the necessity arises'.[8]

So Grove would probably not have recognised himself as an environmental hero. In fact he was quite clear that the future was someone else's problem in this respect. Whilst science could do so little to ease the conditions of the present, he suggested, 'it seems an over-refined sensibility to occupy ourselves with providing means for our descendants in the tenth generation to warm their dwellings or propel their locomotives'.[9] But if not an environmental hero, what about a Welsh one? There has certainly been a flurry of interest in Grove in Wales in recent years – and this book is in part a product of it. A little over a decade ago the Welsh Government commissioned a poll to discover the hundred greatest Welsh heroes. Aneurin Bevan won, closely followed by Owain Glyndŵr, but Grove was there at ninety-one.[10] I suspect that if the poll were repeated now he might do even better. But does it make sense to even think of Grove as Welsh (heroic or otherwise)? Would such an identification tally with his own sense of who and what he was? And if Grove was Welsh, what did (and could) that mean at different points during his lifetime – and what might the identification mean to us in the twenty-first century?

As we shall see, Grove's upbringing and background in industrialising Swansea was to play a crucial role in his formation as a man of science. Even after Grove had established himself on the metropolitan scientific scene, his Swansea connections clearly remained important. There were times when Grove's Welsh identity came to the fore – as in his efforts to bring the BAAS to Swansea and south Wales, and make a success of the meeting. At other times he was happy to describe himself as an Englishman. He shifted identity with occasion and location. His Welshness could be played both to his advantage and his detriment at different points in his career. The main focus of this biographical study will be Grove's career between the mid-1830s and the mid-1850s as he tried to make a life for himself in science. During this prolific and active twenty-year period Grove invented his nitric acid battery and the gas battery, authored what was widely understood at the time to be one of the period's most philosophically significant scientific texts, and played a key role in transforming the Royal Society. Looking at his various – and in the end failed – efforts to make a place for himself in science offers a way not just of understanding Grove and Welsh science, but the changing contours of the scientific culture through which he moved.

1

A SCIENTIFIC PEOPLE

William Robert Grove was born in Swansea on 11 June 1811. His father, John Grove, was a well-heeled and well-connected local merchant whose family had been becoming increasingly influential figures in civic affairs for a number of years, if not decades. John Grove married Anne Bevan on 25 January 1810 in the parish of Llangyfelach on the outskirts of Swansea. The Bevans were themselves an influential local clan whose roots around Llangyfelach and the Gower could be traced back several centuries. The young couple settled at Mount Pleasant in the substantial townhouse called 'The Laurels' owned by John's father, William (and which stood approximately where the central police station is now). They shared Mount Pleasant with some prestigious neighbours. A few houses away was the palatial Georgian mansion, the Willows, where lived Lewis Weston Dillwyn, owner of the Cambrian Pottery and a fellow of the Royal Society. Swansea was growing during these early decades of the new century, and the spread of big, comfortable houses for prosperous men and their families is a reminder that the town was getting wealthier too. An 1818 engraving of Mount Pleasant, drawn by Thomas Baxter (who was employed by Dillwyn at his pottery works), depicts an idyllic scene for the young Grove's upbringing. It was a scene that would shortly disappear though, as Swansea became an increasingly industrial town, dominated by copper as well as the devastating effects of the smelting industry on the landscape.

By the time young William Robert arrived in the summer of 1811, the Groves were clearly well established and influential in the town.

His father, John, was on the governing board of the newly established Swansea Dispensary in 1808, and one of the town's paving commissioners a year later. By the 1830s he was a member of the newly instituted town council and an alderman. At various times during his career John Grove was the portreeve, the town mayor and the deputy lieutenant of Glamorganshire, as well as a Justice of the Peace. One of John Grove's brothers, Thomas, was also a member of the town council and an alderman during the 1830s. Another member of the dynasty, William Grove, John's father, was an alderman in 1802 as well as serving at least three times as the town's portreeve, in 1790, 1800 and 1810. The portreeve was the equivalent of the town mayor before the municipal government was reformed and the borough corporation replaced by a town council during the 1830s. He was also a leading member of the Fund to Obviate the Inconvenience Arising from Copper Smoke, which had been established to investigate ways of reducing the pollution spewed out by the copper smelting that was in the process of transforming Swansea's economic fortunes and its environment.[1] A William Grove was also portreeve in 1818 and 1822, but this may have been his son, and our Grove's uncle and namesake, William Robert.

The information is sketchy, but it is enough to paint a picture of a family that was deeply involved in Swansea's cultural and political life during the early decades of the nineteenth century. The Groves were described as merchants, but it seems clear that they were involved in a wide and varied range of commercial activities in the town and beyond. Amongst other things, they were clearly quite substantial property owners in Swansea itself. At his death in 1896 William Robert Grove owned land and property in the town and elsewhere in south and west Wales. Some of that had presumably been acquired during his own lifetime, but a good proportion was inherited from his father. Besides property, and acting as agents for other landholders, the firm of William Grove & Son dealt in a variety of goods. In 1804 they sold a Boulton & Watt stationary steam engine, for example. A few years later William Grove was the agent responsible for hiring masons for bridge repair at Killay. At various times they sold a malthouse in Cowbridge and even a colliery in Loughor. Unsurprisingly in a maritime town, much of their business was to do with ships. They often acted as agents selling ships' cargo or even the ships themselves. This was a family firmly embedded in the economic life of the town.[2]

As to where the Groves came from, according to a correspondent in the *Cambrian*, writing shortly after William Robert Grove's death in 1896, they were relatively recent imports to Swansea from the nearby Gower peninsula. George Gibbs, the *Cambrian*'s correspondent in the matter, wrote that Grove's 'ancestors lived on a farm in the parish of Reynoldston, Gower, then owned by Thomas M. Talbot, Esq.' After one of the family's sons stepped in to help a Swansea surveyor working for the Talbots' Penrice estate, he so impressed the man with his 'energy and efficiency' that the surveyor, a Mr Hall 'imported him to his office at Swansea, where by dint of persevering attention to business, and good conduct, he rapidly rose to fortune, and laid the basis of that eminence which his descendant attained'.[3] These events are supposed to have occurred 'towards the end of the last century'. If true, they must in fact have taken place somewhat earlier than that (possibly to Grove's grandfather or great grandfather) since the Groves' assured place in Swansea civic culture by the beginning of the nineteenth century suggests that they were not recent arrivals. There is at least one more reference to another Grove as having been descended from 'the old Gower family, from which has sprung the greatest scientific philosopher of our times, namely, Sir William Robert Grove'.[4]

Most of the scanty biographical sources and obituaries that give any account of Grove's upbringing agree that he received his early education at the hands of the Rev. Evan Griffiths. Griffiths was at this time the headmaster of the Swansea Free Grammar School and one or two later accounts describe Grove as having been a pupil. Whether or not this was so seems debatable, however. Having been originally established in 1682 by Hugh Gore, bishop of Waterford and Lismore, the grammar school was in poor shape by the early decades of the nineteenth century, and closed in 1842 following Griffiths's death. A letter in the *Welshman* the previous year complained that the school was effectively moribund, its buildings rented out and the funds used entirely to pay the headmaster's salary rather than providing for the poor pupils who were meant to be its beneficiaries. Grove is more likely to have been one of the 'pay scholars whom he [Griffiths] is at liberty to receive without limitation'.[5] Whatever education Grove received from Griffiths (Latin and Greek only according to the *Welshman*'s correspondent), he was then sent to Bath to be tutored by the Rev. J. Kilvert. The links with Bath and the Kilvert family is intriguing. Lewis Weston Dillwyn, John Grove's near neighbour in Mount Pleasant, was a friend of the Kilverts and might well have been instrumental in the choice of tutor for the young Grove as he prepared for Oxford and Brasenose College.

Grove himself traced his own interest in science to reading at the age of twelve a story in which 'a boy, for the purpose of curing superstition in a younger brother, made an electrical machine, and astonished him with a display of its wonders, and how he invoked the aid of phosphorus to increase the child's astonishment'.[6] Inspired, he started to make his own apparatus from odds and ends like old apothecaries' bottles and syringes and carried out his own experiments. His scientific exploits were sufficiently destructive that 'his father was led to discourage, although he did not actually forbid his scientific pursuits'. These sorts of stories of early experimental precociousness are commonplace and need to be taken with a suitable pinch of salt. But Grove himself noted that his 'grandfather had scientific tastes to some degree', and that his 'grandmother's brother ... was a good amateur chemist and astronomer'.[7] The grandfather was almost certainly the William Grove mentioned earlier who was active in the Fund to Obviate the Inconvenience Arising from Copper Smoke. The 'grandmother's brother' is more difficult to identify. Though the dates do not seem quite right he may have been the engineer Benjamin Hill from Clydach, mentioned in some obituaries as an early scientific acquaintance and who was noted locally as a chemist and astronomer as Grove describes him.[8] The Kilvert connection in Bath may be relevant in this context as well, since the Kilvert family were certainly active in scientific circles there.[9]

Given John Grove's (alleged) dislike for his son's scientific pursuits, it is worth pausing briefly to consider the young William Robert's education and prospects in a family context. One source suggests that Grove was 'intended by his father for the Church' but that 'conscientious scruples interfered with the father's desires' leading to the decision to aim for a career in law instead.[10] Grove was almost certainly the first of his family to acquire a university education and it is worth speculating as to why his father felt one to be a useful acquisition for his son, or why a career in the Church was considered at all. The more obvious path for a prosperous Swansea merchant's son to follow would have been in his father's footsteps. Sending a son to Oxford was an important step for someone like John Grove, and should be understood as a mark of substantial dynastic ambition on his part. This rather looks like an attempt to signal a decisive break from a commercial past to a professional and gentlemanly future. William Robert Grove's arrival at Brasenose College Oxford in the autumn of 1830 therefore marked a potential new beginning for the family as a whole. It was a new beginning that was nevertheless still firmly rooted in the broader civic ambitions of Swansea itself.

Swansea Science

There can be little doubt of the ambitions that Swansea's civic leaders – and John Grove presumably among them – had for their town during the opening decades of the nineteenth century. At the end of the previous century, Swansea was a comparatively small seaside town with a population of about six thousand. It had been a thriving port since Tudor times, at least, and during the second half of the eighteenth century had developed a growing reputation as a tourist resort. Well-heeled English travellers came to enjoy the scenery and the recent fashion for sea bathing. The town's inhabitants were accustomed to supplement their incomes by renting out rooms to these moneyed visitors. By the standards of Welsh towns at the beginning of the nineteenth century, 6,000 was in fact quite a substantial population. Swansea was certainly a bigger and more prosperous town than Cardiff, its rival port thirty miles or so further east along the coast of the Bristol Channel, for example. As far as many of its inhabitants were concerned Swansea clearly was the pre-eminent town in Wales. If anywhere merited the title of the Welsh metropolis, it was their town.[11]

By the beginning of the nineteenth century, Swansea was as much an industrial centre as it was a port and seaside resort. Since the 1720s the town had been steadily growing as an important centre for the copper smelting industry. Swansea's proximity to the copper mining districts of Cornwall on the other side of the Bristol Channel, with the port allowing easy import of the ore and export of the finished product, along with the availability of coal mined in the Swansea valley made the town ideal for the purpose. The town also had a thriving pottery industry. Copper smelting in particular had a huge impact on the town and the environment. The Fund to Obviate the Inconvenience Arising from Copper Smoke with which William Grove senior had been involved was an effort to find ways of reducing this impact without overly reducing the profits flowing into the copper kings' pockets. The wealth that flowed into Swansea through the copper trade would play a large part in the town's expansion and improvement during the nineteenth century's opening decades, after all. The role copper played in the local economy also meant that Swansea was a town where the sorts of practical chemical skills associated with smelting were highly valued. It was a town that understood how science might be made useful.

Swansea's civic leaders during the opening decades of the nineteenth century, whilst Grove was growing up, were intent on improvement. The town corporation took on plans to widen and pave streets and improve sanitation. In 1808 they passed a resolution

> that it will be highly advantageous to the town and borough of Swansea to obtain an Act for paving, repairing, cleansing, lighting and watching the streets and other public passages and placed within the said town and borough, and for removing and preventing nuisances and obstructions within.[12]

Local gentry joined together to fund a new theatre and assembly rooms, setting up the Swansea Tontine Society for that purpose. The tontine's subscribers included individuals from as far afield as Bath and Gloucester – an indication of the increasingly important economic and cultural role that the town played not just across south Wales but for the Welsh marches and south-west England as well. There was plenty there to feed the intellect too. Swansea had two circulating libraries by the beginning of the nineteenth century. By 1808 the town had a public subscription library whose proprietors included a list of local dignitaries such as Lewis Weston Dillwyn. As well as libraries, Swansea also had a flourishing book trade and in 1807 saw the publication of the *Cambrian*, describing itself as 'the first and only newspaper published in Wales'.[13]

The circles in which the Grove family moved during young William Robert's childhood must have included the leading lights of Swansea society. Lewis Weston Dillwyn was a close family neighbour. John Grove's involvement in Swansea's civic affairs would certainly also have brought him into contact with Dillwyn as well as other notables such as John Henry Vivian, both of whom were with Grove senior members of the first town council in 1836. Vivian was a fellow deputy lieutenant of the county. Both these men were prominent industrialists, the Dillwyn family having made their fortunes through potteries whilst the Vivians were copper smelters. Significantly, both Dillwyn and Vivian were fellows of the Royal Society who had wide-ranging scientific interests as well as an extensive circle of scientific acquaintances. Vivian was the first Member of Parliament elected for Swansea following the introduction of the Reform Act in 1832. He remained as MP until his death in 1855, when he was replaced by Lewis Llewellyn Dillwyn, Lewis Weston Dillwyn's son. Lewis Weston Dillwyn himself was the first reform MP for the county of Glamorgan. The Vivians, unsurprisingly given their copper interests, were relatively recent immigrants to the town from Truro in Cornwall. The Dillwyns were also quite recent arrivals. Lewis Weston Dillwyn's father was a Philadelphia Quaker who had returned to Britain during the war of independence and settled in Essex.

Whilst there is little to show what the political colours of the Grove dynasty might have been, the Dillwyns and the Vivians were reform-minded and liberal in their politics. In this they were fairly typical of the rising class of prosperous, often nonconformist industrial men that emerged out of the Industrial Revolution and who came to dominate public life in newly industrialising towns and cities not only in south Wales but in Britain more generally. An interest in science was a common feature of this new industrial middle class. The first Literary & Philosophical Institute was established in Manchester in 1781 and was followed by others in Bath, Birmingham, Leeds, Liverpool and Newcastle, amongst others. Societies like these offered their members a steady diet of scientific lectures by local enthusiasts as well as visitors from further afield. Their libraries stocked the latest scientific journals. This was the kind of thing that Swansea's scientific enthusiasts had in mind when they took steps towards the end of 1835 to establish the Swansea Scientific & Literary Institution. The new institution's founding members included Dillwyn and Vivian, of course. But they also included a young William Robert Grove, who had completed his studies at Oxford in 1833 and entered Lincoln's Inn in London to study for the bar, as well as his father John.

Swansea's was certainly not the first scientific society in south Wales and it is not surprising that such societies proliferated there during the first half of the nineteenth century. South Wales was exactly the sort of nonconformist, industrialising territory that was fertile ground for such activities and organisations. From the middle of the eighteenth century, at least, natural philosophy had started to become an increasingly central aspect of the rising professional and industrial middle class's cultural worldview. There is clear evidence that literary and scientific societies were being established in towns and cities across Wales from the 1830s onwards as well as indications of earlier gatherings. It is possible to follow the tracks of travelling scientific lecturers across the country from newspaper reports of their activities. Local scientific societies also offered their own members opportunities to demonstrate their prowess as natural historians, electrical experimenters or chemists. Not only the scientific bigwigs like Dillwyn or Vivian, who operated on a national stage, but enthusiasts with more modest pretensions had their chance in the limelight at their meetings. Local dignitaries donated collections of scientific instruments, botanical or geological specimens, or books for libraries. In Wales, as elsewhere, scientific societies were symbols of a town's civic aspirations and claims to culture.

One role played by these scientific societies was as neutral territory where men from different political camps could make common cause.

When the Neath Museum & Society for Promoting the Arts & Sciences was established early in 1835, the *Merthyr Guardian* commented that it 'is always pleasing to see men of various political opinions joining hand and heart in the noble object of the cultivation of the arts and sciences, and the intellectual and moral improvement of those around them'.[14] The society aimed to 'form a depository for specimens in the various branches of Natural history', said Swansea's *Cambrian* newspaper. Local worthy Howel Gwyn (later a Tory politician and High Sheriff of Glamorganshire) was appointed president and banker John Rowland the treasurer.[15] Neath's promoters of science clearly had some heavyweight backing, succeeding in attracting William Conybeare, dean of Llandaff, as a patron. When Conybeare delivered a lecture on local geology to the society later that year, his performance attracted a crowd of 'about four hundred of the leading families and the respectable inhabitants of the town of Neath and the western parts of the county'.[16] Two years later, they succeeded in attracting George Bidell Airy, the Astronomer Royal, to perform for them – another sign of how local Welsh enthusiasts for science often had links with broader scientific networks.[17]

When the Merthyr Literary & Scientific Institution was established in 1837, its founding committee included amongst its members a Baptist minister, a local doctor, the solicitor William Meyrick and the radical and soon-to-be local Chartist leader Morgan Williams. The secretaries were Edward Lewis Richards and Taliesin Williams. The list reads like a gathering of local reform-minded moderates and radicals of just the kind that had joined forces from the 1820s onwards to campaign for parliamentary reform. These were also men who would have been on different sides of the barricades a few years before they rallied around the banner of science, during the Merthyr rising in 1831. Meyrick, as William Crawshay's solicitor, was prominent in local politics in defence of his patrons' interests and contested Merthyr against Guest on a pro-Church Tory ticket in 1834. Morgan Williams had been involved in radical politics from an early age and was co-editor of a radical newspaper, *Y Gweithiwr/The Workman*. He had also been instrumental in persuading many of Crawshay's adherents to defect to Guest's camp in 1834, forcing Meyrick's withdrawal from the contest.

Taliesin Williams (son of the prominent antiquarian, self-proclaimed Jacobin and inventor of the Gorsedd, Iolo Morganwg) was himself a local schoolmaster, Unitarian and moderate defender of unions and workmen's rights. The other secretary, Edward Lewis Richards, came with similar credentials.

The rhetoric deployed to unite these uneasy bedfellows around the proposed scientific institution is significant. The cause that was supposed to bring the warring sects together was one of civic and national unity under the banner of useful knowledge. Addressing the meeting, Edward Lewis Richards offered an interesting blend of civic insecurity, confidence and ambition.

> When we for a moment reflect ... on the position of Merthyr Tydfil, its industry, energy and mercantile importance ... it really appears a matter of astonishment as well as of regret, that so many years should have elapsed without some attempt in the centre of this place at forming so valuable, so desirable an institution.

Richards demanded that his cheering audience should

> look to our neighbouring towns, and observe the efforts daily made to inculcate by every means varied and useful knowledge. Let us do this, and with the blush of shame confess that Merthyr alone hangs back in the great progress of civilization. – Swansea, with a population of 15,000 people, Carmarthen, with a population of 10,000, Neath, with only 5,000, even Newbridge, now emerging from the obscurity of a country village, have either literary, scientific or mechanic's institutions: and all classes unite in commendation of their value and usefulness.

It was a scandal that 'Merthyr, the largest iron work in the world, a model for nations in its energy and enterprise' lacked such an institution, he said.[18]

As Richards noted, Swansea by now had indeed acquired its own scientific society. Inspired by Neath's example, the town's leading citizens had come together to establish the Swansea Literary and Scientific Institution. The new society's first lecture meeting was held on Friday, 11 December 1835, when G. G. Bird discussed the structure of the eye, and called in the inaugural address which preceded his lecture for 'each individual who has it in his power' in the town, to 'run quickly and light his torch at the sacred temple of learning, and come hither in his turn and hold it forth for the use of others'.[19] It soon became clear, too, that Swansea's enthusiasts for science had extremely high ambitions for their fledgling institution. Early meetings were held at rooms held in Castle Square, but it was not long before they decided that something more impressive was required. On 24 August 1838, Lewis Weston Dillwyn, the institution's president, recorded in his diary how he

> drove as President to a large public Breakfast, given by the Building Committee, and about 2, walked in procession and laid the first stone of the Royal Institution. It was accompanied by a Royal Salute of 21 guns, the ringing of bells, the cheers of a vast multitude and a prayer by Dr. Hewson.[20]

As it had in the case of Merthyr Tydfil, the banner of science raised in Swansea could provide an uncontentious focus for civic pride in uncertain political times. Like Merthyr, though to a lesser degree, Swansea had suffered from its own discontents throughout the first half of the nineteenth century as industrialisation and urban expansion transformed the area. There were riots over food shortages in 1801 and William Grove the elder as portreeve had been instrumental in dealing with them. In 1823 the corporation voted an additional fifty guineas for the local police in view of the continued unrest. Only a few years after the Swansea Lit & Phil was established, and just as the plans to transform it into the Royal Institution of South Wales were being laid, the Rebecca riots convulsed Swansea and the surrounding countryside. As civic leaders and guardians of the law, many of the Royal Institution's prominent men were at the forefront of the campaign to quell the uprising, as we shall see later. Occasions like laying the Royal Institution's foundation stone offered opportunities for displays of unity and played to shared aspirations about Swansea's place and ambitions for the future. They were tangible reminders to the population of what their leaders wanted as they worked to transform the town in their own image.

Making the Man of Science

William Robert Grove may well himself have been part of the 'vast multitude' that gathered to watch that foundation stone being laid. He was after all a member of both the renamed institution's council and its building committee. He had clearly been associated with the Swansea Literary and Scientific Institution right from the very beginning. He was a member of the institution's first committee, along with a number of other Swansea luminaries. It seems likely that the young Grove was one of the new institution's instigators and one of its leading lights during its early years. If so, it put him in prestigious company. As well as the Groves' neighbour Dillwyn as president, the council included John Henry Vivian and Dillwyn's son John Dillwyn Llewelyn amongst the vice presidents. Another vice president was Christopher Rice Mansel Talbot, a member of the powerful Talbot landowning clan, founder of extensive ironworks in nearby Aberavan and builder of Margam Castle. Like Dillwyn and Vivian, he was a fellow of the Royal Society. William Henry Fox Talbot, the pioneer of photography with whom Grove would have much to do in the future, belonged to another branch of the clan. Another founder member was Henry de la Beche, newly appointed as director of the Ordnance Geological Survey in 1832. It was the Geological Survey's activities that had brought de la Beche to Swansea. It was a sign of the young Grove's ambitions and his self-assurance that he was confident enough of himself to mix in circles like these – and that such powerful players in the game of science accepted him as one of themselves.

Grove had left Oxford in 1832 with only an ordinary degree. In other words he had elected not to sit the examinations that would lead to honours. This was not at all an uncommon decision. Relatively few students at either Cambridge or Oxford proceeded to honours during this period. Whatever Grove thought of his official studies, there would certainly have been enough scientific activity in Oxford during the early 1830s to pique the interests of a young man whose thoughts were turning to natural philosophy. One of his tutors was Baden Powell, the university's Savilian Professor of Geometry, for example. Powell played an active part in debates about the place of natural philosophy within the university.[21] Grove might have attended the lectures delivered at the Ashmolean Museum by Stephen Peter Rigaud, the university's reader in experimental philosophy. Another prominent Oxford natural philosopher – and another one with Swansea connections – was William Buckland. Buckland, famous for his attempt in *Reliquiae Deluvianae* to demonstrate that the story of Noah's flood was vindicated by the geological record, would have been well known to Grove as the discoverer of the Red Lady of Paviland, the ancient skeleton found in a cave on the Gower peninsula in 1823. There is no record that he attended the meeting, and he had already left Oxford by the summer of 1832, but Grove must have been aware that the second meeting of the British Association for the Advancement of Science (BAAS), founded a year earlier in York, was to meet at Oxford that year.[22]

We know little else of Grove's activities at Oxford, or of what he did in the years between leaving university and the establishment of the Swansea Literary and Scientific Institution. He entered Lincoln's Inn and was called to the Bar in 1835. Several contemporary biographical sources suggest ill health and continental travel, in the course of which a taste for science was rekindled, and there is no particular reason to doubt that anecdotal evidence. He had no formal scientific training, but that was the rule, rather than the exception, until much later in the century. But it seems likely that natural philosophical pursuits had formed some part of his experience from his Swansea childhood and through his student career. Such activities and interests certainly formed an important part of the Swansea circles in which he and his family moved during those formative years. He must also have done something, at least, to make the likes of Dillwyn, Talbot and Vivian take him seriously. His contributions to the Swansea society were not just confined to sitting on committees either. During the 1836–7 season Grove offered the institution a couple of lectures on optics. The following season he delivered another lecture on galvanism. It seems likely that, if nothing else, he had been spending some of his time in scientific self education.

The lectures' contents offer some glimpse of his early interests. The *Cambrian* certainly reckoned his first performance a success. They thought 'that Mr. Grove's explanations, illustrations, and diagrams, were peculiarly effective' and described how 'after having gone through the subject of lenses reflectors &c., the lecture was concluded by the Drummond light (which Mr. Byers prepared for the occasion) being passed round the Hall by a parabolic reflector, which was most brilliant'.[23] His second lecture, a week later, dealt with physical optics and ranged from discussions of Newton's experiments on the constitution of white light ('with the aid of coloured diagrams'), through accounts of telescopes and David Brewster's theory of the solar spectrum, to a comparison of the competing undulatory and corpuscular theories of light. The 25-year-old philosopher had no qualms about choosing sides. He was a supporter of the undulatory theory. Grove cited a number of factors in favour of the wave theory of light, and offered an explanation of the phenomenon of Newton's rings that avoided Newton's own 'somewhat unphilosophical theory of fits of reflexion and transmission'.[24] Grove had been doing his homework, and was doing more here than simply repeating textbook platitudes. He was finding a place for himself on the scientific map and starting to lay the groundwork for his own individual philosophical position.

Grove's lecture on galvanism to the now renamed Royal Institution of South Wales the following year was even more intriguing in this respect. He gave his audience (which was 'numerous') a brief survey of galvanism's history, with an account of Luigi Galvani's experiments in animal electricity, followed by a description of his rival philosopher Alessandro Volta's researches that led to the invention of the voltaic pile in 1800. This was all preliminary to a detailed discussion of the chemical theory of the voltaic pile's action, and descriptions of different kinds of battery. Grove treated his audience to an account of the Cruickshank and Wollaston voltaic batteries and highlighted the basic problem that dogged these early batteries in that 'the action sinks after the first few minutes of immersion'. He described de la Rue and Daniell's efforts to remedy the problem before proceeding to describe some experiments and batteries of his own. He demonstrated various kinds of batteries that used only a single metal (unlike conventional batteries that used two – usually copper and zinc), including an 'economical battery of Mr. Grove's invention, made of alternate plates of iron and thin wood, such as that used by hatters'. He finished it all off with an account of electro-magnetism, throwing in the 'original conjecture, that, in addition to many other natural phenomena, Saturn's ring, which rotates in its own plane, may be an electro-magnetic effect'.[25]

Grove was now clearly experimenting, and experimenting well. Even more explicitly than his previous excursions on optics, this lecture was original work, not reheated textbook matter. In the years since leaving Swansea for Oxford, Grove had transformed himself into an accomplished natural philosopher and experimenter. He had acquired the necessary skills of the electrician and clearly felt himself the equal of men such as John Frederic Daniell, professor of chemistry at Kings College London. He was experimenting with himself too, and creating a public scientific persona. If newspaper reports are to be credited, he was an adept and effective performer. Hints of some of the concerns that would preoccupy him throughout his career as an experimenter and philosopher were already visible. There was the concern with efficiency and utility in that 'economical battery' made of easily available (in Swansea at least) bits and pieces, for example. It was a position that probably went down quite well amongst the hard-nosed Swansea industrialists and professional men that would have made up the bulk of his audience at the Royal Institution of South Wales. Grove knew exactly what appealed to such men. He had grown up amongst them. By 1839 he was starting to spread his wings beyond Swansea though. A few months after his appearance at the Royal Institution of South Wales, the *Cambrian*, Swansea's local newspaper, gave extensive treatment to a paper by 'Mr. Grove (of Swansea)' at the BAAS, in which he discussed a new battery of his invention. This was the nitric acid battery that would make Grove's name as an experimenter on the international stage. It was, the *Cambrian* reported the chemist Thomas Graham as saying, 'an important improvement'.[26]

So by the time that the Royal Institution of South Wales's impressive building opened its doors during the summer of 1840, Grove was an active contributor to Swansea science. He was an important contributor to the new institution's activities both as a member of various committees and as a productive experimenter and lecturer. He was on an equal footing with its leading members. Like them too, he had developed a network of scientific relationships that stretched well beyond Swansea and south Wales, as we shall see in more detail in the next chapter. His performance at the Birmingham meeting of the BAAS was only one example of this. Grove's local reputation as a man of science increasingly came to depend on his activities and his reputation elsewhere, just as much as it depended on his performances in the institution's lecture room. The Swansea connection would remain an important feature of his scientific life throughout the subsequent decade nevertheless. The town and the family home at the Laurels provided him with an occasional base and the friendships made locally with other active scientific men like Lewis Weston Dillwyn and his son John Dillwyn Llewellyn were useful on the wider stage as well. By the mid-1840s, Grove would be a prominent and increasingly influential member of the metropolitan philosophical community, but his links to Swansea and south Wales would continue to play a vital role in sustaining his reputation.

2

THE METROPOLIS OF SCIENCE

Of course, by the time that Grove unveiled his nitric acid battery before his audience at the chemical section of the British Association for the Advancement of Science's (BAAS) Nottingham meeting in 1839 he was already in the process of leaving his Swansea youth behind him. Following the end of his Oxford studies Grove entered Lincoln's Inn in London in November 1831 to start the process of training to become a barrister. He was called to the Bar four years later in 1835. During those four years Grove presumably divided his time between the family home in Swansea and London lodgings. He would have been required to keep term at Lincoln's Inn in order to qualify. Very little is known about Grove's activities during these early years in London. Several biographical sources suggest that he suffered from ill health during this period and that it was during this period of enforced leisure that his interest in experiment was rekindled.[1] If that is so, then London was certainly a good place to be for a young man keen to cultivate science. The metropolis at the beginning of the Victorian age was home to an impressive range of scientific institutions of all kinds, catering to all tastes and classes.[2] This was the capital city of a growing empire that depended on commerce and industry. Many of its citizens thought science offered both rational entertainment and the prospect of economic gain. The apprentice barrister could take his pick.

London's premier scientific institution when the young Grove arrived there at the beginning of the 1830s was the Royal Institution. Described with acerbic wit by the essayist Thomas Carlyle as 'a kind of sublime Mechanics' Institute for the upper classes', the Royal Institution had been established in 1799 by émigré American loyalist and mercenary soldier, Benjamin Thompson.[3] Located on fashionable Albemarle Street, it was plain to what audience the Royal Institution would appeal. First under Sir Humphry Davy and then with Michael Faraday as director of the laboratory, the Royal Institution made itself into a centre for scientific lecturing that catered to the requirements of polite society. When he was appointed director of the laboratory in 1825, Faraday carried out a thorough overhaul of the institution's activities, introducing new features such as the Friday Evening Discourses and the annual Christmas Lectures as well as reorganizing the regular lectures. This was all aimed at increasing the Royal Institution's appeal to a well-heeled clientele whose public lives were shaped by the rhythms of the London season. Fashionable London duly thronged to the Friday Evening Discourses in their hundreds, to listen to Faraday, or a favoured invited speaker, expound on the latest science. As the novelist George Eliot would remark decades later, looking back at this period in the institution's history, good society 'gets its science done by Faraday'.[4] The Royal Institution was where they got it.

The mechanics institutes to which Carlyle compared the Albemarle Street institution were quite emphatically not aimed at catering for the needs of the fashionables. The London Mechanics Institute had been founded in 1824 following a campaign in the *Mechanics Magazine* to establish an institution where the metropolis's working men could educate themselves in the sciences. Whilst the *Mechanics Magazine*'s editors, Thomas Hodgkin and Joseph Robertson, argued that the new institution should be run by working men with their own resources and for their own benefit, moderate reformers such as George Birkbeck took a different view. The moderates won the argument and by the 1830s there was little left of the original radical vision. The membership of mechanics institutes tended to be drawn from the lower end of the middle classes rather than from amongst the skilled working men that had been the original target. By the 1830s the London Mechanics Institute had an imposing building for itself in Southampton Buildings on Chancery Lane, a stone's throw from where Grove was receiving his preparation for the law at Lincoln's Inn. Its members had access to reading rooms and a library as well as a variety of lectures on different scientific subjects.[5]

London also had its share of literary and philosophical institutions of the kind that had become popular during the final years of the eighteenth century. The Surrey Institution, founded in 1807 and modelled in part on the Royal Institution, had foundered several years before Grove arrived in London in the early 1830s. The Russell Literary & Scientific Institution was established a year later in 1808 and occupied a substantial building on Great Coram Street near Russell Square. Unlike the Surrey, the Russell flourished. It offered members an extensive library for their annual subscriptions of one guinea, as well as regular courses of lectures. A little more substantial than either the short-lived Surrey, or even the more successful Russell, was the London Institution, established in 1806. By the 1830s it occupied impressive premises on Finsbury Circus on the fringes of the City. As well as these establishments, there were a number of smaller societies such as the London Literary Institution in Aldersgate Street and the Western Literary Institution in Leicester Square that catered for those with scientific interests. Scientific lectures could be found at places like the Society of Arts too. The presence of the Royal Society and the more recently established specialist societies such as the Astronomical and Geological Societies added to the metropolis's thriving scientific culture.

A young man like Grove looking for scientific entertainment and edification in London during the 1830s would have found himself wandering through a marketplace for knowledge. The number and variety of scientific institutions meant that the metropolis offered real opportunities for men hoping to make (or at least supplement) their living by scientific lecturing. Lecturers like the German émigré Frederick Accum, J. T. Cooper and William Sturgeon did the rounds of these institutions, offering courses of lectures and individual performances. Some, like Accum, offered private courses in natural philosophy in their own homes or in rented premises. It was a highly competitive market and lecturers needed to tailor their performances accordingly to appeal to their intended audiences.[6] Some of these institutions – notably the Royal Institution – offered permanent employment. There, Michael

Faraday had a position that not only offered him security but provided him with the resources that made his extensive experimental researches in electricity and magnetism possible. In turn these scientific institutions and individuals depended on networks of instrument-makers and craftsmen who provided them with the material resources they needed for their performances. Lecturers did not just stand in front of their audiences and talk. They performed experiments, showed specimens and colourful maps and diagrams, or deployed a magic lantern to illustrate their arguments. These sorts of lectures were as much about spectacle as they were about science.[7]

Scientific spectacle was very much to the fore at venues like the National Gallery of Practical Science (or Adelaide Gallery) that opened its doors to the public in 1832, or its competitor the Royal Polytechnic Institution that opened a few years later. At places like these visitors paid a shilling at the door to gain entry into a hall of scientific wonders.[8] The Adelaide Gallery was located in the Lowther Arcade on the Strand, opposite the scientific instrument-maker Edward Marmaduke Clarke's Laboratory of Science. The Royal Polytechnic Institution was on Regent Street – another site designed to attract the attention of curious (and moneyed) passers-by. At the Adelaide Gallery, visitors could expect to see 'clever professors ... teaching elaborate science in lectures of twenty minutes each'.[9] At the Polytechnic were 'half-globes, brass pillars, and water troughs so charged with electricity as nearly to dislocate the arms of those that touched them'.[10] They offered another venue where instrument-makers and inventors could show off their productions, and where lecturers like William Sturgeon and J. T. Cooper could perform public experiments. Within months of Faraday's discovery of electromagnetic induction, the Adelaide Gallery's resident instrument-maker, Joseph Saxton, had devised an apparatus to produce a spark by induction. Within a year his magneto-electric machine was on show there.[11]

The science of electricity played well at these sorts of venues devoted to scientific spectacle. It was at the Adelaide Gallery and then at the Royal Polytechnic Institution that the short-lived Electrical Society of London held its meetings from 1836 onwards. This group of electrical enthusiasts represented a different kind of scientific network from the elite circles of the Royal, or even the London or Russell Institutions. Leading figures in the society included the popular lecturer and instrument-maker William Sturgeon, the instrument-maker Edward Marmaduke Clarke and William Leithead, superintendent of the Department of Natural Magic at the Coliseum – one of the city's many sites for exhibition and spectacle.[12] The Electrical Society held regular meetings at the Adelaide Gallery, availing themselves of the facilities there for experiment. Their interests lay primarily in the more practical aspects of experimentation. They demonstrated new kinds of batteries and electromagnetic engines. It was at a meeting of the Electrical Society of London that James Prescott Joule first presented his own experimental efforts at investigating the efficiency of such devices.[13] Given that many of its members were active participants in the London lecturing network it should be no surprise that developing electrical demonstrations and displays was a particular concern. William Sturgeon, for example, had first made a name for himself a little over a decade before the society's foundation as the inventor of the electromagnet – a device to magnify the magnetic effects of electricity so that they could be seen properly by a lecture audience.[14]

As Grove arrived in London, intent on pursuing his scientific interests, one of the decisions he had to make was about where he fitted on this complex and interlocking cultural map. The experience of science – what science meant and how it was understood by performers and audiences alike – differed significantly depending on the places where it was performed. Science as performed by Michael Faraday to his comfortable audience at the Royal Institution looked quite different from science at the Adelaide Gallery or the Royal Polytechnic Institution. This was not just a matter of taste. Some radically different ideas about what science was, what kinds of people its practitioners should be, and what sort of science was appropriate for different social classes, were implicit (and sometimes explicit) in these various forms of scientific organisation.[15] We do not know to what extent Grove, when he first arrived in London, envisaged himself as a producer as much as a consumer of scientific knowledge. It seems a fair assumption that during these early years in the metropolis he was reading books and attending lectures.

By the time he appeared as a lecturer before his fellow members of the Royal Institution of South Wales he was already fashioning himself as a scientific authority. The kinds of societies he joined and the place he occupied in London's scientific networks would have a great deal to do with deciding just what kind of scientific authority he wanted to be.

Making a Scientific Man

Most biographical sources agree that Grove became a member of the Royal Institution in 1835. Coming as he did from a prosperous provincial background Grove would certainly have had both the financial wherewithal and the social clout to be a member of this elite institution. It was a privilege that did not come cheap and that certainly carried expectations in terms of status. The Royal Institution was a scientific club for gentlemen (and ladies). If Grove really was trying to explore the opportunities that London offered for establishing himself as a man of science rather than a mere spectator, then membership of the Royal Institution could prove to be a vital asset. Furthermore, Grove already had some of the resources he needed to take advantage of the Royal Institution's opportunities. His relationship with Swansea neighbours such as Lewis Weston Dillwyn and Henry Vivian – both fellows of the Royal Society who operated on a national as much as a local scientific stage – offered him an additional advantage. Their patronage could pay dividends in terms of entry to more exclusive scientific circles than might otherwise have been open to him. But by cultivating the attention of the sorts of gentlemen of science who could be encountered at the Royal Institution Grove was also setting out his own stall as a practitioner. He was going to be a gentleman of science too.

Whilst cultivating scientific networks at the Royal Institution Grove was also pursuing other relationships as he continued the process of leaving Swansea behind and settling in the metropolis. On 27 May 1837 William Robert Grove married Emma Maria Powles at Stamford Hill Chapel in London.[16] Emma was the daughter of Danish-born financier and speculator John Diston Powles, who had made his fortune through judicious (and occasionally controversial) investments in South

America. Where and how Grove and Emma had become acquainted remains unknown. There seems to be no record that John Diston Powles was a member of the Royal Institution, for example.[17] The fact that Grove married at the relatively early age of twenty-five suggests that his legal career was already flourishing, as well as underlining his own faith in his capacities for the future. Middle-class men during this period did not as a rule marry until they were capable of providing for a family – though in Grove's case his father's wealth, and that of his prospective father-in-law may have made that need less urgent. The young couple seem to have spent much of the following year abroad, maybe on an extended honeymoon. Their first son, Florence Crauford Grove (later a pioneering mountaineer and founder member of the Alpine Club) was born in Florence on 12 March 1838.

Whatever the reasons for Grove's foreign travel during this period, he was clearly taking advantage of the opportunities his peregrinations allowed for furthering his reputation as a man of science. Some of his early publications on electricity were sent to the *Philosophical Magazine* from Paris, and others were communicated to the Académie des Sciences there. Either through contacts in Swansea or at the Royal Institution he had secured an introduction to the French electrical experimenter Antoine César Becquerel who passed on his communications to the Académie. His first communication to the *Philosophical Magazine*, dated 26 October 1838, was addressed from Swansea, however. Grove and family were presumably ensconced with his parents at the Laurels. The battery Grove described in this letter was very similar to the one he would demonstrate to the Royal Institution of South Wales a few months later. It was made from 'common stout millboard' formed into a trough and 'covered with a thin layer of cement', into which a number of 'four-inch squares of common iron' and 'unglazed porcelain plates of the same dimensions' were placed 'at about three tenths inch distance'. When a 'solution of sulphate of copper and dilute acid' was

> poured into the alternate cells, a very active series is formed by the precipitation of copper on one surface of the iron; that which I

formed was of twenty plates: the shock, without coils or condenser of any description, was so powerful as to be scarcely tolerable.[18]

In his first appearance in the pages of a prestigious philosophical journal Grove was showing himself to be a seasoned experimenter (and he was presumably pleased to see that his own contribution followed immediately after one from Michael Faraday). The sophisticated reader casting an eye over Grove's letter would have realised at once that it had been written by someone familiar with the latest discussions of electrochemical theories as they applied to the practical construction of voltaic cells. But his contribution would have struck a chord with members of the Electrical Society of London too, with their interests in the practicalities of utility and display. Someone like William Sturgeon, for example, would have been particularly struck by Grove's closing suggestion that the

> advantage of this form, where series and sustained power are required, I consider to be its extreme œconomy, a single cheap metal being employed instead of two expensive ones; the greater durability of iron as compared to zinc; the cutting into squares, so that none is wasted; and the tiresome process of soldering being altogether dispensed with.[19]

This was the sort of virtuous parsimony that might well have appealed to readers back home too, reading the local boy's first excursion into print.

The emphasis on economy was clear in Grove's second published outing too. Dated from Swansea a couple of months later, this letter to the *Philosophical Magazine* described continuing research on the topic of the 'œconomical constant battery' he had described the last time. In these experiments Grove was not just putting what he called his 'unimetal' battery through its paces, he was also trying to use it as a tool to investigate how batteries more generally worked and how their behaviour might be improved. The abiding fault of electric batteries since Volta invented his pile in 1800 was their tendency to run down quite quickly. They could not be counted on to generate a constant current for any length of time. A few years earlier in 1835 John Frederic Daniell, professor of chemistry at King's College London, had put forward a design for a constant battery. Grove's plan was to build a better one. This time his experiments were with iron and with copper. He described how with

> twelve plates acidulated water was rapidly decomposed: with a pair of copper plates each exposing about 36 square inches of surface, a Ritchie's rotating magnet was whirled rapidly round, exhibiting small but brilliant sparks; its revolution continued for several hours without the addition of fresh acid.[20]

It is worth emphasising the extent to which some of the experiments described in this communication were very similar to the ones that Grove would use to illustrate his lecture to the Royal Institution of South Wales on galvanism a few months later. As much as anything else, these experiments seem to have constituted preparations for public performance, and may be a sign that Grove was already contemplating (or at least dreaming about) the prospect of such a role for himself. The number of individuals who successfully made a living through scientific lecturing during the first half of the nineteenth century, still less by occupying a permanent position, was very small indeed. Nevertheless, Grove may already have been developing aspirations for just such a career. It certainly looks as if he were intent on being more than just an occasional experimenter. He was developing his own distinctive programme of research and taking pains to make his scientific vocation public. A few months later a further brief communication from Grove appeared in the pages of the *Philosophical Magazine*, discussing the action of zinc (a common component of electric batteries) in acidulated water. It was October before the magazine published an account of Grove's new nitric acid battery, which he had demonstrated both at the Académie des Sciences and the BAAS annual meeting. By the time this account appeared Grove had further improved the battery, and could boast that he had 'underrated very considerably the power of the battery' in his presentation in Birmingham.[21] Whilst all this was going on Grove had also become a father again. His second son, Coleridge, was born on 26 September 1839.

If Grove had hoped that his announcement of a powerful new battery would prove decisive in his efforts to establish a scientific reputation he was not to be disappointed. His demonstration of the battery at the BAAS was well received and opened up a number of new opportunities for Grove to show off his scientific credentials. The account of his new battery published in the annual *Reports of the British Association for the Advancement of Science* as well as his communication to the *Philosophical Magazine* were republished in the prestigious *Annalen der Physik*, edited by the German natural philosopher Johann Christian Poggendorff, then one of the leading European journals of natural philosophy. More immediately and (in the short term at least) more importantly, Grove received an invitation early in 1840 from Michael Faraday to deliver one of the Royal Institution's Friday Evening Discourses. Established by Faraday in 1827 these regular events were by now well established as some of the more prestigious in London's scientific season. The invitation certainly offered Grove a chance to display his flair for experiment and demonstration before an audience packed with potential patrons. He took full advantage of the opportunity. He put the nitric acid battery through its paces, producing 'an arc of flame, of an inch and a quarter long' between charcoal points, and showing how 'the blade of a pruning knife, which Faraday would submit to the test, was consumed to the handle in an instant'. He could boast how news had just arrived that Professor Jacobi in St Petersburg had sailed a boat on the river Neva 'with forty-eight of these combinations at the rate of two miles and a half an hour'.[22]

Grove's deft handling of his invention and its public appearances had two important consequences that were to prove decisive for his future career. On 26 November 1840 he was elected a fellow of the Royal Society. Fellowship at this time was largely a matter of acquiring sufficient backers to sign the nomination (Grove was to be instrumental in changing this state of affairs a few years later) and the list of Grove's supporters underlines the extent to which he had been successful in making himself known in the right circles. The list included Grove's old

Swansea patron John Henry Vivian, but it also included Henry Moseley and Charles Wheatstone, the professors of natural and experimental philosophy at King's College London, Thomas Graham, the professor of chemistry at University College London, and Richard Phillips, one of the editors of the *Philosophical Magazine*. The second decisive event was Grove's appointment, early in 1841, as professor of experimental philosophy at the London Institution. The committee established to consider the chair had included, amongst others, the institution's president Thomas Baring, the prosperous wine merchant and enthusiast for electrical experiment John Peter Gassiot. Correspondence between Grove and Gassiot makes it clear that it was the latter who was instrumental in putting Grove's name forward for the position.

The London Institution

William Robert Grove's trajectory from unknown provincial electrical enthusiast to fellow of the Royal Society and professor of experimental philosophy at the London Institution had been quite spectacular. As professor at the London Institution he had his own laboratory, as well as the funds to purchase apparatus, and an assistant to work under his supervision. The assistant was George Thomas Fisher, a former medical student, who would be paid £1 6s. a week to help prepare experiments for the lectures and keep a laboratory diary. The laboratory was 'a large quadrangular apartment completely fitted with chemical furnaces and apparatus of the most approved construction'. The institution also provided him with a ready-made and public forum that he could use over the following years to further enhance his own professional and philosophical reputation as a man of science. The lecture theatre was a particularly impressive space. It was 'nearly semi-circular, the inner curve measuring 117 feet 4 inches, and the chord 62 feet 9 inches; the area being sufficient to contain nearly 700 auditors'.[23] Grove had clearly been very lucky in succeeding to ensconce himself in one of the very few permanent and paid positions that the metropolis offered to aspiring natural philosophers – even if the salary of £100 a year was hardly generous. This was luck that Grove had made for himself, nevertheless.

It was not just the power of his new battery that had landed this plum position in his lap. It was the result of his assiduous efforts at building social networks too.

John Peter Gassiot, who, as one of the London Institution's managers, would prove to be a crucial ally for Grove, seems to have been one of his earlier new acquaintances in London as well. The first item of surviving correspondence between the two is dated 17 August 1840 and it is clear from that letter that they already knew each other well enough for Grove to have expressed to Gassiot his frustration that his work as a barrister was interfering with his scientific vocation – and for Gassiot to tantalizingly hold up the prospect of a scientific post for his protégé. Gassiot was wealthy and well connected within the metropolitan scientific elite.[24] Unusually though, he had links lower down the social scale of London science as well. He was the treasurer of the Electrical Society of London too. It may well have been that rather odd combination that drew him to the young Grove. Perhaps he recognised a similar ambivalence. Perhaps he recognised that Grove's combination of theoretical ambition and practical concern matched the requirements of the London Institution too. Grove was duly invited to perform at one of the soirées that the London Institution had established in imitation of the Royal Institution's Friday Evening Discourses on 20 January 1841 and a few months later in April commenced his duties as professor.

It is safe to assume that the fact that John Diston Powles, Grove's father-in-law, was also one of the London Institution's proprietors, as the institution's members were known, helped smooth his path to the professorship as well. Powles's links with the London Institution underscores the establishment's avowed relationship with city and financial interests. The London Institution had been founded by money men like Powles, who wanted a scientific club more responsive to their concerns than the aristocratic Royal Institution.[25] In his inaugural address at the laying of the new institution's foundation stone, the city lawyer Charles Butler reminded his doubtless appreciative audience, that 'science and commerce are mutually dependant: each assists the other, and each receives from the other, a liberal return'.[26] The combination of science and commerce could 'record the heavens, delve the depths of the earth, and fill every climate that encourages them with industry, energy, wealth, honour, and happiness. – To civilization, to virtue, to religion, they open every climate; they land on every shore; they spread them on every territory.'[27]

This theme of the interrelationship of science, industry and imperialism was just the thread that would run through Grove's own inaugural address to the London Institution when he took up his chair there a quarter of a century later, too. In fact, comparing the two it is easy to imagine that Grove must have had Butler's lecture on his desk before him as he prepared his own. Delivering a lecture 'at the first Soirée for the season' on the progress of the physical sciences during the previous year was one of the duties that his new professorship carried with it. Even though the London Institution had by 1841 been in existence for the best part of four decades it had not previously appointed a professor, so it was decided that for his first performance Grove would deliver an overview of the progress of the physical sciences since the London Institution's foundation. The lecture took place on Wednesday, 19 January 1842, presumably at the London Institution's palatial lecture theatre. This was Grove's opportunity to demonstrate to the members that they were getting their money's worth. Like their equivalents at the Royal Institution, Grove's audience here wanted to be entertained through science. But they also wanted to be assured that this was an entertainment (and an entertainer) that could manage philosophy whilst still understanding on what side the bread was buttered.

What Grove offered this audience was a finely tuned account of how progress in science and progress in imperial commerce could not be disentangled. He tickled their vanity by reminding them of the role places like the London Institution played in making London 'the metropolis of the civilized world' and how the science such institutions promoted had already 'wrought epochal changes in our knowledge, and will work gradual changes in our political history'.[28] Grove romped through the divisions of the physical sciences, telling his audience, for example, how the steam engine demonstrated 'the grandest mastery of mind over matter', and how 'the most recondite scientific discoveries find, ere long, their application to the arts, and add, not only to the mental advancement, but also to the comforts, the luxuries, and the power of man'.[29] Advances in electricity meant that they now had the means 'to fuse the most intractable metals, to propel the vessel or the carriage, to imitate without manual labour the most costly fabrics, and in the communication of ideas, almost to annihilate time and space'. If anyone had dared suggest any of this at the beginning of the century, he scoffed, 'the prophet, Cassandra-like, would have been laughed to scorn'.[30]

All the examples of scientific progress that Grove chose – from Jean-Victor Poncelet's water-wheels, through Charles Babbage's calculating engine, to electromagnetic engines, telegraphs, electrotypes and photography – were instances of natural forces taking on the job of displacing human labour. This was an account of progress that put science at the heart of industry – just where his listeners wanted it to be. 'Why is England a great nation?' Grove asked them.

> Is it because her sons are brave? No, for so are the savage denizens of Polynesia: She is great because their bravery is fortified by discipline, and discipline is the offshoot of Science. Why is England a great nation? She is great because she excels in Agriculture, in Manufactures, in Commerce. What is Agriculture without Chemistry? What Manufactures without Mechanics? What Commerce without Navigation? What Navigation without Astronomy?[31]

The picture he drew for his audience was one that put the London Institution right at the heart of the matter of England. It was also, quite strategically, a picture that placed the view of science that Grove endorsed right at the heart of the matter as well. If this was Grove's manifesto, he had chosen the right place and the right audience for its announcement.

What the published version of the lecture misses, of course, is that this was a performance as much as a text. Grove himself noted in the preface to the printed text, 'printed by order of the mangers of the London Institution, at the request of the proprietors', that 'the Lecture was delivered extempore' and could not be taken as more than a copy from memory of the original. That had included illustrations (and experimental demonstrations, presumably). The institution's lecture theatre was designed specifically for that kind of visual extravaganza. There was, for example a 'small room' behind the theatre itself 'connected with an opening over the head of the Lecturer, intended for the Exhibition of Illustrative Transparencies' (and, incidentally, phantasmagoria performances).[32] The *Literary Gazette*, at any rate, was impressed by Grove's performance. 'When we say that the several subjects enumerated above were treated in a masterly manner', their correspondent reported at the end of his account of the lecture, 'and that the peroration was eloquent and impressive, some idea of the gratification afforded to the numbers assembled may be formed'. They noted that the 'large theatre was thronged – and subsequent to the discourse crowds surrounded the novelties exhibited in the library'.[33]

Experimenting with a Career

Grove's position as professor of experimental philosophy at the London Institution was, more than anything else, a source of status. As the incumbent of one of the metropolis's very few salaried positions in science (even if the salary was rather less than what would be needed for a gentleman of Grove's background to maintain a household in any comfort), his was now a voice to be taken seriously by his peers. To many struggling hopefuls trying to make their way through science in the metropolis, his must have seemed an enviable position. Grove now had a platform and the experimental resources that he needed to carry on with his researches. The London Institution's laboratory seems to have been well stocked with the usual apparatus needed for lecture demonstration. The institution's managers had made sure to acquire everything they thought a competitor to the Royal Institution might require when the new building was completed in 1819. Since then, they had made sure the collection of apparatus was kept up to date. One item of apparatus no longer available to Grove was the institution's famous 'Great Battery', built by William Haseldine Pepys and consisting of

2,000 plates of zinc and copper, since by the 1830s it had 'been virtually destroyed by frequent use, from the repeated action of the solvents employed'.[34] During his first year as professor, Grove spent just under £50 on laboratory apparatus. The following year he spent almost £80.[35]

We can get some sense of what kinds of things Grove was doing in his new laboratory from a report presented by Charles Vincent Walker to the Electrical Society of London on 16 September 1841. Walker gave the society an account of a visit to Grove's laboratory and presented a table of experimental results that Grove had showed him. If this is anything to go by, Grove had been carrying out a detailed and exhaustive comparison of the powers of different kinds of voltaic combinations. He had tabulated the deflection of a galvanometer needle produced by various batteries. As Walker emphasised, 'Mr. Grove did not give me this table under the impression that the degrees of deflection can convey to the mind any real idea of the relative *amount* of superiority, possessed by one arrangement over another.'[36] He was simply tabulating how different batteries performed this particular task. This was a perspective that would have been perfectly familiar to the Electrical Society audience. This was how battery makers typically proceeded. They demonstrated their batteries' powers by using it to produce a range of displays – like deflecting a galvanometer needle, or decomposing water, or heating a length of wire. This was not, strictly speaking, a matter of measurement. It was a matter of finding out what battery was best for different kinds of display.[37]

That Grove was producing this sort of table suggests that one of the things he was doing in his London Institution laboratory was carrying on with the investigations of battery performance that he had been conducting for the past several years and that had led to the invention of his nitric acid battery. Sometime during 1842 he also seems to have taken another look at those experiments that he had first detailed in a postscript to his letter to the *Philosophical Magazine*'s editors back in December 1838. In these experiments he had noticed that electricity seemed to be generated when plates of platinum were sealed in tubes containing oxygen and hydrogen respectively. He hoped, he said, 'by repeating this experiment in series, to effect decomposition of water by means of its composition'.[38] Grove was referring to the common method of registering the presence of a flow of electricity by the decomposition of water. Wires from both poles of a battery would be placed inside upended glass tubes standing in a trough of water. When electricity passed through, oxygen gas was released from the water at one pole, and hydrogen gas at the other. The amount of gas produced this way was often used by electricians such as the members of the Electrical Society to demonstrate the relative power of different voltaic combinations.

On 22 October, Grove dashed off a note to Michael Faraday, describing his experiments and inviting him to visit the London Institution's laboratory to see this 'curious voltaic pile … composed of alternate tubes of oxygen & hydrogen'.[39] He told him how the arrangement gave him 'an unpleasant shock' as well as decomposing both potassium of iodide and water. Gassiot, he mentioned, had already seen it. The first public announcement of the new battery came a couple of months later, in a note to the editor of the *Literary Gazette*. The letter gives the impression that Grove was responding to a request from the editor for more information about the invention, so it is possible that William Jerdan, then the *Gazette*'s editor, had already seen the strange new battery in action in Grove's laboratory. Grove explained how 'when the alternate tubes are filled with oxygen and hydrogen, there are notable voltaic results, spark, shock, divergence of gold leaves, deflexion of galvanometer, decomposition of water and electrolytes'.[40] The decomposition of water was, he said, a 'beautiful instance of the correlation of natural forces'. What Grove had in mind here was the way in which the gases in the battery generated electricity as they combined into water, whilst the water in the voltameter decomposed to produce oxygen and hydrogen when electricity passed through it. It was the symmetry he found intriguing.

Just as Grove had considered his 'unimetal' battery as a tool for getting at the relationship between chemical and electrical action, so he thought this peculiar new combination offered a way of getting inside the operations of the voltaic cell. That certainly seems to be largely what he thought about this new piece of apparatus he called the gas battery as he first started tinkering with it. Whilst most electricians were interested in simply generating new technologies of display, Grove was interested in dissecting them too. The same sort of approach was there in his researches into the electric spark as well. In an account of some experiments on the voltaic discharge sent to the *Philosophical Magazine* on 7 May 1840 he described his attempts to investigate how the surrounding medium affected the electric spark, for example.[41] These sorts of experiments could have practical implications too. In 1845 Grove published an account in the *Philosophical Magazine* describing how electricity could be used to provide safe illumination in coal mines. This, he said, was why he had been interested in investigating the voltaic spark. In the end he settled instead on using coils of thin platinum wire heated until they glowed as a source of illumination. You could read your newspaper by it, he declared.[42] A few years later an account of the lamp was included in the report compiled by Henry de la Beche and Lyon Playfair on gases and explosions in collieries.[43]

Not all Grove's work in his new laboratory was electrical. In 1844 he delivered a lecture on photography to the chemical section at the annual meeting of the BAAS, held that year in York. It was a subject in which his Swansea acquaintance, John Dillwyn Llewelyn (Lewis Weston Dillwyn's son) – married to Henry Fox Talbot's cousin – was already developing a keen interest. The experiments Grove described at the York gathering were an attempt to simplify the calotype process that Fox Talbot had patented by 'obtaining a paper capable of giving positive photographs in one process, and avoiding the necessity of transfer'.[44] Grove had in fact been dabbling with photography a few years earlier too – this time combined with electricity. An account of a 'voltaic process for etching daguerreotype places' was communicated on his behalf by Gassiot to a meeting of the Electrical Society on 17 August 1842. The experiments were 'a remarkable instance ... of the effects of the imponderable upon the ponderable: thus, instead of a plate being inscribed as 'drawn by Landseer, and engraved by Cousins', it would be 'drawn by Light, and engraved by Electricity!'[45]

Grove's first few years at the London Institution were highly productive. Between 1841 and 1845 he published an impressive fourteen papers, including a couple of hefty contributions to the Royal Society's

FIGURE 1 William Robert Grove, *Portraits of Men of Eminence in Literature, Science, and Art, with Biographical Memoirs, the Photographs from Life*, by Ernest Edwards, BA (London: Alfred William Bennett, 1865)

Philosophical Transactions. He was clearly in the process of building a substantial reputation for himself not only on the national, but on the international scientific stage, as his papers were translated and reprinted in prestigious continental publications like Poggendorf's *Annalen* and his friend Auguste de la Rive's *Bibliotheque Universelle*. Nevertheless, beneath the surface, Grove's position was not quite as secure as it may have appeared to casual observers. Grove was furious when a communication he had read at a meeting of the Royal Society in 1841 was rejected for publication in the prestigious *Philosophical Transactions*. He sniffed corruption and nepotism in the air and scribbled an angry letter to Faraday, fulminating that 'this paper even in an improved form ... would also have been rejected unless I had *made interest* for its insertion & this I will not do'.[46] He also found himself embroiled in a nasty little spat with John Fredric Daniell, the professor of chemistry at King's College London, about the originality of his nitric acid battery. It seems clear that Grove's efforts at entering the higher echelons of London's scientific elite did not meet with entirely unalloyed delight on the part of all its members.[47]

3

THE CORRELATION OF PHYSICAL FORCES

On a spring evening towards the end of April 1842, Grove's friend and patron, the wealthy wine merchant John Peter Gassiot, organised a soirée at his home in Clapham Common. This was not an ordinary soirée – it was an Electrical Soirée, and guests were invited to marvel at a whole range of electrical wonders. The event featured amongst the exhibits 'one hundred series of Professor Grove's nitric-acid battery, which brilliantly ignited about nine feet of iron wire', as well as 'electro-types, electro-tints, electro-chromes, electro-plating, &c. &c.' There was also an electromagnetic machine, designed by Charles Wheatstone which, said the *Literary Gazette*'s correspondent attending the evening, 'attracted our particular attention and admiration'.[1] Gassiot hosted a similar soirée the following year, to celebrate the visit to London of the eminent Swiss natural philosopher and electrician Auguste de la Rive (whom Grove knew from his continental travels). Grove's battery was the *pièce de résistance* again, this time used to produce a powerfully bright electric spark. The light was so bright that to 'look at it direct was painful ... it penetrated the outer darkness, shooting over the lawn; but now softened into the sweetest moonlight, and yet clothing the shrubs and turf with intense green'.[2]

Events like these offer a nice illustration of the sort of electrical world in which Grove found himself in 1840s London. Electrical experiments were often geared to the production of spectacular effects. The same year that Gassiot's guests wondered at the brilliant illumination generated by Grove's batteries, visitors to the Royal Polytechnic Institution on Regent Street marvelled at the effects produced there by the wonderful hydro-electric machine exhibited by William George Armstrong (later famous as a pioneering gun manufacturer and ennobled as the first baron Armstrong of Cragside).

> The passage of electricity over the tinfoil on the tubes was far more brilliant, and the aurora borealis exceeded in intensity and beauty anything we had ever witnessed; the violet colour was brighter, and at the same time deeper, and the exhausted received showed more plainly the progress of the electric spark

gushed the *Morning Chronicle*.[3] Utility as well as beauty was a matter for display, hence the demonstrations of electromagnetic engines and electric lights at Gassiot's soirées. Useful machines not only demonstrated electricity's importance for economic progress but acted as useful lessons about the place of electricity in the economy of nature. They helped show that electricity – quite literally according to experimenters like William Sturgeon – made the world go round.

New devices like the electromagnetic telegraph invented in 1837 by Charles Wheatstone and his business partner William Fothergill Cooke were things to marvel at as well as offer hopes of a tidy profit to investors. The telegraph seemed to annihilate time. It held out the promise that messages that had once taken days to deliver from one end of the country to the other could now be transmitted instantaneously. Edward Copleston, the bishop of Llandaff, was so excited following his visit to Wheatstone's laboratory to see the telegraph in action that he rhapsodised enthusiastically in his diary about how it 'far exceeds even the feats of pretended magic and the wildest fictions of the East ... a thousand times more than what all the preternatural powers of which men have dreamt of and wished to obtain were ever imagined capable of doing'.[4] He even dreamed about it. Wheatstone and Cooke, along with rival telegraph promoters like Edward Davy, put their instruments on show in places like the Adelaide Gallery and the Royal Polytechnic Institution in their efforts to attract investors and astound the public.

'You did not expect to have a son turned showman', quipped Davy in a letter to his father, but he recognised it as an essential part of the business of electrical invention.

The same could be said about the many attempts to develop a reliable electromagnetic engine for locomotion. Machines like these featured prominently in the experimental efforts of members of the London Electrical Society. William Sturgeon, for example, invented an electromagnetic engine in 1832 and put it on show at the Adelaide Gallery. A few years later the entrepreneurial American inventor Thomas Davenport put his electromagnetic engine on show at the Adelaide Gallery too. The press were ecstatic at its possibilities, one correspondent enthusing that

> I cannot discover any good reason why the power may not be obtained and employed in sufficient abundance for any machinery – why it should not supercede steam, to which it is infinitely preferable … Half a barrel of blue vitriol, and a hogshead or two of water, would send a ship from New York to Liverpool; and no accident could possibly happen, beyond the breaking of the machinery, which is so simple that any damage could be repaired in half a day.[5]

Of course, it was Jacobi's successful demonstration of his electromagnetic engine to power a boat on the Neva in St Petersburg, using Grove's nitric acid battery that had played a role in bringing Grove's invention to public notice too.

Electrometallurgy was another industrial product of this electrical enthusiasm for utility and spectacle. Electricity could be used to cover cutlery, plates and ornaments made from cheap metal with a thin layer of silver – just the thing for the emerging middle classes to buy to show off their growing prosperity. This was all about spectacle too. As one electrical writer put it,

> a person may enter a room by a door having finger plates of the most costly device, made by the agency of the electric fluid. The walls

of the room may be covered with engravings, printed from plates originally etched by galvanism, and multiplied by the same fluid. The chimney piece may be covered in ornaments made in a similar manner. At dinner the plates may have devices given by electrotype engravings, and his salt spoons gilt by the galvanic fluid.[6]

Newspapers, magazines and specialist journals like the *Annals of Electricity*, edited by William Sturgeon, or the *Electrical Magazine*, edited by the railway (and later telegraph) engineer Charles Vincent Walker were full of optimism about the electrical future and vivid descriptions of the fantastic spectacles of nature generated through electrical experiments. This was the world to which Grove's own electrical researches during the early part of the 1840s, as he worked to make his name at the London Institution, belonged.

In fact, much of Grove's experimental work at the London Institution seems to have been concerned with investigating electrical spectacle and the instruments made to put electricity on display. In a contribution to the *Philosophical Magazine* in 1843, building on the observation that 'in certain (probably in all) cases of the development of a voltaic current a reaction was induced by the voltaic force itself, and that upon the cessation of the initial force the reacting force was apparent in an opposed direction', Grove proposed a novel method for increasing battery power by getting the reacting force to act in the same direction as the original force. He was confident that the technique 'appears to me to promise results of some practical value; the œconomy of this method of applying force is evident, we get all but a double product with a single consumption'.[7] He was also continuing his investigations on voltaic ignition – producing sparks and heating wires to incandescence – the kinds of experiments that produced the electric arc light put on show at Gassiot's soirée, or the electrical miners' safety lamp he would demonstrate a couple of years later.[8] One thread running through this work was his interest in showing how electricity could be used to produce a whole range of different effects involving heat, light, magnetism and so on, and how those effects could be put to use to produce results both useful and spectacular.

This was the kind of experimental work that Grove placed before the Royal Society in his Bakerian Lecture delivered on 19 November 1846, not long after he had resigned his position as professor at the London Institution. Delivering the Bakerian Lecture was an important honour, and Grove's choice of topic was significant. What he gave the assembled fellows was an account of how bits and pieces of the technology of display could be turned to analytical purposes. He had devised an instrument for analysing gases – an eudiometer – that

> possesses the advantage of enabling the operator either to detonate or slowly to combine the gases, by using different powers of battery, by interposing resisting wires, or by manipulation alone, – a practised hand being able by changing the intervals of contact to combine or detonate the gas at will.[9]

He described experiments that compared the heating and illuminating effects of a wire connected to a battery and enveloped in different gases – and reminding his audience of the experiments he had carried out to investigate the possibility of using electricity as a source of light in mines. Grove concluded the lecture with an experiment that used his apparatus to directly decompose water into oxygen and hydrogen by heating it, first with a current carrying wire, and then, to be sure it was not an electrical effect, using a blowpipe flame. The Bakerian performance was an attempt at synthesising different aspects of Grove's experimental life into a single and coherent whole.

Uniting Nature

Unity was a prevailing concern for many early Victorian natural philosophers – the unity of nature and the unity of science itself. They worried that natural philosophy seemed to be splintering into a spectrum of specialisms, both institutionally and intellectually. Dissatisfaction with the Royal Society had led to the establishment of rival organisations devoted to particular sciences, of which more in the next chapter. There were a number of competing views about just what science was, how it should be done and by whom. Corresponding to these various visions of science were a range of visions of the unity of nature too. Different natural philosophers offered their own big pictures of a uniform nature governed by an underlying principle or power. The idea of progress was central to many of these scientific visions. Eighteenth-century natural philosophers like Joseph Priestley had offered unified visions too, but theirs tended to emphasise nature in a state of equilibrium as opposing powers balanced each other. Nineteenth-century visions tended to portray nature as progressive instead. Nature did not stay the same – it had a past and a future. It developed rather than remaining static. Ideas like these about the *nature* of the natural order were, as everyone agreed, the key to understanding the social world as well.[10]

John Herschel's *Preliminary Discourse on the Study of Natural Philosophy* was published in 1831 as the first volume of the *Cabinet Cyclopaedia*, edited by Dionysius Lardner, professor of natural philosophy at the London University. The series was designed to introduce the different branches of natural philosophy to its readers, with Herschel's volume providing a synoptic overview of the whole field. Man, said Herschel, 'is constituted a speculative being; he contemplates the world, and the objects around him, not with a passive, indifferent gaze ... but as a system disposed with order and design'.[11] For a natural philosopher, taught

> to trace the operation of general causes, and the exemplification of general laws, in circumstances where the uninformed and unenquiring eye perceives neither novelty nor beauty, he walks in the midst of wonders: every object which falls in his way elucidates some principle, affords some instruction, and impresses him with a sense of harmony and order.[12]

Herschel argued that the unfolding of the laws that underpinned the progressive operations of nature offered a way to improve the mind and convince the unbiased observer that a higher intelligence must lie behind it all. The natural philosopher's task was to unravel the relationships of cause and effect, and the 'first great agent which the analysis of natural phenomena offers to our consideration, more frequently and prominently than any other, is force'.[13]

Herschel's attempt to synthesise science was a huge success. Another widely read overview with an emphasis on the unity of the sciences was Mary Somerville's *Connexion of the Physical Sciences*, published a few years later in 1834. As she put it at the book's beginning:

> The progress of modern science, especially within the last five years, has been remarkable for a tendency to simplify the laws of nature, and to unite detached branches by general principles. In some cases identity has been proved where there appeared to be nothing in common, as in the electric and magnetic influences; in others, as that of light and heat, such analogies have been pointed out as to justify the expectation, that they will ultimately be referred to the same agent: and in all there exists such a bond of union, that proficiency cannot be attained in any one without a knowledge of others.[14]

She took her readers on a guided tour of the sciences, starting with astronomy, the forms and movements of celestial bodies, gravitation, light, heat, electricity and magnetism, before returning to gravity and the diffusion of matter throughout space. She explained how

> innumerable instances might be given in illustration of the immediate connection of the physical sciences, most of which are united still more closely by the common bond of analysis, which is daily extending its empire, and will ultimately embrace almost every subject in nature in its formulas.[15]

William Whewell, soon to be master of Trinity College in Cambridge, took a different tack. He found the unity of the sciences in their history. His door-stopping three volume *History of the Inductive Sciences*, published in 1837, traced the development of the sciences 'from the earliest to the present times', as the book's subtitle had it. According to Whewell, what the history of science showed was that science was progressive, and that the progressive development of its different branches shared a common historical structure. The 'history of each science forms a whole in itself, divided into distinct but connected members, by the *Epochs* of its successive advances', he argued.[16] The history of each particular discipline was presented as one of progress from an initial prelude, during which the fundamental ideas that would form the science's intellectual core emerged into clarity and prominence, and would end with 'the discovery which marks the epoch, seized and fixed for ever the truth which had till then been obscurely and doubtfully discerned'. The epoch would be followed by a sequel, 'during which the discovery has acquired a more perfect certainty and a more complete development among the leaders of the advance'.[17] The end of all this progress was a mature, permanent science characterised by its key concepts and ideas.

Whewell and Herschel's Cambridge contemporary Charles Babbage took a very different view of the unity of nature and science. His *Ninth Bridgewater Treatise*, published the same year as Whewell's *History*, was a robust response to the series of Bridgewater treatises published by the Royal Society to demonstrate the conformity of science and religion. For his model of the operations of universal natural laws Babbage took the calculating engine that had been one of his main preoccupations for the best part of two decades by 1837 when his book was published. He used the example of his calculating engine to show how nature could evolve over time whilst still following the same overarching laws. In just the same way that Babbage could adjust his calculating engine so that it followed different arithmetical rules at different points during its operation, an intelligent creator could have created a progressive evolving nature, with new laws coming into operation at different points in its evolution. Such a view could even explain miracles as part of a natural order of things. By this view,

> no motion impressed by natural causes, or by human agency, is ever obliterated. The ripple on the ocean's surface caused by a gentle breeze, or the still water which marks the more immediate track of a ponderous vessel gliding with scarcely expanded sails over its bosom, are equally indelible.[18]

The anonymous *Vestiges of the Natural History of Creation*, published in 1844, offered another, and far more controversial, attempt at offering an overarching view that brought the different branches of natural philosophy back together again. *Vestiges* was an early Victorian bestseller.[19] Its author (later revealed to be the Edinburgh publisher Robert Chambers) offered his readers a comprehensive history of the universe revealed through the operation of a progressive natural law. He started with the universe's origins and explained the creation of stars and planets by means of the nebular hypothesis, which argued that they emerged from huge clouds of nebular gas in space as the clouds started rotating and coalescing into solid spheres. He suggested that life on Earth had its origins in spontaneous generation brought about by electricity. 'Electricity we also see to be universal', he declared, and 'if it be a principle concerned in life and in mental action, as science strongly suggests, life and mental action must everywhere be of one general character'. Electricity was therefore the lynchpin that linked progress in nature with mental and with social progress. It was a grand theory that drew on the latest natural philosophical discoveries to make bold claims about the origins of the universe, the nature of life and the future of society.[20]

Vestiges infuriated many of Grove's fellow metropolitan gentlemen of science, like Herschel, Whewell and Babbage. They thought that its uses of their science were a misuse and that the whole thing brought science into disrepute. There were some other attempts at forging grand and ambitious theories of everything that made even *Vestiges* seem tame. In 1841 the committed follower of the utopian socialist Robert Owen, Thomas Simmons Mackintosh, published his *Electrical Theory of the Universe*, with its grand claim that everything in the universe could be reduced to electricity. From this he concluded that it was 'better to view man as an organized machine'. From that point of view 'every motion, every process, is progressing towards a point in which it will terminate; and life is a process which only exists by a continual approach towards death. Eternal life and perpetual motion are almost, or altogether, synonymous.'[21] This was the kind of unrestrained speculation that demonstrated to some of Grove's fellow men of science how dangerous it could be to allow the wrong kind of people to theorise about the deeper meaning of their experiments. The task of telling the public what their experiments meant could only be entrusted to very particular kinds of people.

Correlation

When Grove sat down sometime in 1846 to write *On the Correlation of Physical Forces*, he must have had some of these concerns at the front of his mind. The furore that had surrounded the publication of *Vestiges* and the savaging it and its anonymous author had received at the hands of some of his fellow men of science was only a few months past. At the very least, by writing such a book he was making a very strong claim about who he thought he was as a natural philosopher. The urgency of his claim is underlined by the fact that the immediate occasion for putting pen to paper was his resignation as professor of experimental philosophy at the London Institution. He was relinquishing one source of status and authority within the metropolis's elite philosophical community. Was publishing *Correlation* an effort to lay claim to another? According to the title page, the essay was 'the substance of a course of lectures delivered in the London Institution in the year 1843', and Grove acknowledged in the preface that his materials were 'much better adapted to oral than to written explanation; they demand at every step experimental illustration, and suggest questions which the Lecturer has an opportunity of answering, but which the writer cannot anticipate'.[22] But as well as offering a convenient hook from which to hang his experimental demonstrations, the concept of correlation offered Grove a way of displaying serious philosophical credentials and presenting his own vision of progressive nature.

Grove introduced his essay with a discussion of the concept of causality and the aims of experimental philosophy. Like most of his contemporaries (both Herschel and Whewell introduced their books like this too) he pointed towards the English philosopher Francis Bacon as the originator of a properly empirical scientific method. Unlike Herschel, for example, though, he was critical of the notion that natural philosophy should be a search for essential causes. He thought it was better thought of as a 'search after facts and relations' instead.[23] Causation might be fine in theory, but not in practice, he argued. You might be able to assign a specific cause to a specific effect on a particular occasion, but you could not then argue that that cause was always the cause of that effect. The same effect might be produced by some different cause on another occasion. Grove raised the relationship of electricity and magnetism as an example of the sort of thing he meant. John Herschel, for one, had suggested that electricity was the cause of magnetism, but following Faraday's experiments on magneto-electricity it was clear that under different circumstances magnetism could cause electricity, So 'if electricity causes magnetism, and magnetism causes electricity, then electricity causes electricity, which is absurd'.[24]

This was why Grove thought is made more sense to talk about correlation. He gave a detailed description of just what he meant:

> The position which I seek to establish in this essay is, that the various imponderable agencies, or the affections of matter, which constitute the main objects of experimental physics, viz. Heat, Light, Electricity, Magnetism, Chemical Affinity, and Motion, are all Correlative, or have a reciprocal dependence. That neither taken abstractedly can be said to be the essential or proximate cause of the others, but that either may, as a force, produce or be convertible into the other, this heat may mediately or immediately produce electricity, electricity may produce heat; and so of the rest.[25]

Grove had made a similar point before, in the lecture on the progress of the physical sciences that he had delivered as his inaugural performance at the London Institution five years earlier. There he had argued that

> cause and effect, therefore, in their abstract relation to these forces, are words solely of convenience: we are totally unacquainted with the ultimate generating power of each and all of them, and probably ever shall remain so ... we must humbly refer their causation to one omnipresent influence, and content

ourselves with studying their effects and developing by experiment their mutual relations.[26]

Having established his philosophical framework, Grove went on to make it clear to his readers that correlation was meant to be a way of talking about experiments. It was all to do with experimental practice and demonstration, both in the laboratory and in the lecture theatre. Correlation offered a way of bringing experiments together and making their relationships clear to his audience without offering any philosophical hostages to fortune. It was a way of following force and its transformations around from experiment to experiment. At the same time, of course, the experiments themselves could be offered up as powerful instances of correlation in action. They were a way of making the correlation of forces visible. In a discussion reminiscent of some of Charles Babbage's remarks in the *Ninth Bridgewater Treatise*, Grove argued that 'force cannot be annihilated, but is merely subdivided or altered in direction or character'.[27] 'It is true,' he said, 'that, at a certain point, we lose all means of detecting the motion, from its minute subdivision, which defies our most delicate means of appreciation, but we can indefinitely extend our power of detecting it, accordingly as we confine its direction or increase the delicacy of our examination.'[28] To all intents and purposes, what correlation offered was a way of following those sorts of transformations through experiment.

In the bulk of the essay, Grove went through the various 'affections of matter' that he had listed at the beginning (light, heat, electricity, magnetism, etc.), offering experimental examples of how the relationships between each of them could be understood in terms of the overarching principle of correlation. In this he seems to have been following quite closely the strategy he had followed in the original lectures. The *Philosophical Magazine*'s account of the performances described how in 'each lecture one of the above forces was taken as the initial force or starting-point, and it was shown experimentally how the others were produced by it'.[29] Most of the examples he offered were quite conventional and drawn from well-known experimental demonstrations. To illustrate the case of light, however, he described an experiment of his own devising which offers rather a good example of what Grove meant by correlation, as well as underlining the theatricality of the concept. A daguerreotype plate was enclosed in a glass-fronted box filled with water. A gridiron of silver wire was placed in the box between the daguerreotype plate and the glass front and connected to the plate by a wire to form a circuit in which a galvanometer and a Breuget's helix were also placed in series. Initially, the glass front was covered by a shutter to prevent any light from passing through it, but when the shutter was removed and the daguerreotype plate exposed to light the needles on the galvanometer and the helix were deflected, 'this, Light, being the initiating force, we get chemical action on the plate, electricity circulating through the wires, magnetism in the coil [of the galvanometer], heat in the helix, and motion in the needles'.[30]

Correlation offered a way of fitting Grove's own experimental work into a broader framework of philosophical discovery as well. The main significance of his invention of the gas battery, as far as Grove himself

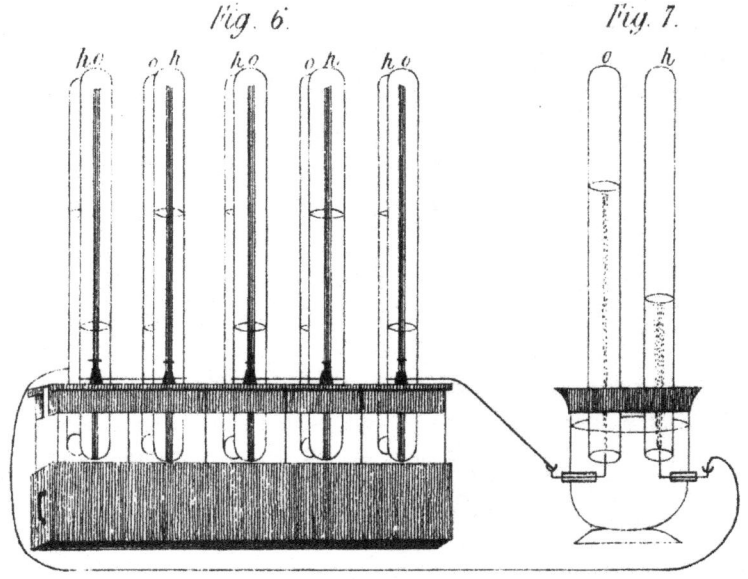

FIGURE 2 The Gas Battery, *Philosophical Transactions*, 1843.

was concerned, for example, was that it offered 'such a beautiful instance of the correlation of natural forces'.[31] Its philosophical value lay in the fact that it offered a way of making correlation visible. It is possible to think of many of his other experimental contributions in the same sort of way too. The nitric acid battery can be regarded as a technology that converted chemical into electrical forces; the work on the decomposition of water by heat that formed part of the subject matter for his Bakerian Lecture to the Royal Society could be understood as identifying a direct correlation between the forces of heat and chemical affinity; his experiments on the direct production of heat by magnetism demonstrated an immediate correlation between the forces of heat and magnetism, and so on. Correlation provided a neat and tidy way of slotting all these experiments into a useful theoretical framework that focused on effects and their relationships. It offered a way of packaging his experimental programme as a coherent whole.

But just what did Grove mean by correlation? He offered a comprehensive definition towards the end of the essay, reminding readers who had just digested pages of detailed experimental procedures of his philosophical credentials. Correlation, he said, 'strictly interpreted, means a necessary mutual and reciprocal dependence of two ideas, inseparable even in mental conception: thus the idea of height cannot exist without involving the idea of its correlate, depth, – the idea of parent cannot exist without involving the idea of offspring'. In the sense that Grove used it in his essay, correlation meant

> a reciprocal production or convertibility; in other words, that any force capable of producing or being convertible into another may, in turn, be produced by it, – nay, more, can itself be resisted by the force it produces, in proportion to the energy of such production, as action is ever accompanied and resisted by reaction; thus the action of the electro-magnetic machine is reacted upon by the magneto-electricity developed by its actions.[32]

There was no point in searching for essential causes, he argued. The idea made no philosophical sense. Natural philosophers needed to acknowledge that 'an essential cause is unattainable. – Causation, is the will, Creation, the act, of GOD.'[33]

Grove's account of cause and effect placed him firmly in a philosophical tradition that would have been very familiar to many of his audience at the London Institution and to the readers of the published essay. The common sense philosophy from which Grove was drawing had its roots in the mid-eighteenth-century attempts to get around the sceptical conclusions drawn by David Hume in his efforts to get to grips with the relationship of cause and effect. There were no rational grounds for supposing that effect would always follow cause, said Hume. Only habit and custom made us think so. The common sense philosophers, led by Thomas Reid, argued against this that our own intuitive common sense and everyday experience were enough to provide secure grounds for belief.[34] Grove had started his assault on traditional notions of cause and effect with the account given by Thomas Brown, student of Dugald Stewart who had himself been one of Reid's disciples. Brown defined a cause as 'that which immediately precedes any change, and which, existing at any time in similar circumstances, has been always, and will be always, immediately followed by a similar change'.[35] What Grove wanted to argue was that his notion of correlation made this kind of abstract reasoning irrelevant. All the philosopher had to do was to follow the forces around, and illuminate their transformations.

Grove was both agreeing and disagreeing with Thomas Brown. He agreed that all that might sensibly say about causes was that they seemed ingrained in the way people talked about the world. He disagreed because the thought that correlation offered a better vocabulary for the business. Just as with the common sense view of cause and effect, correlation was both a feature of nature and a feature of science. It described the way in which the different forces of nature were related to each other, and provided a way of talking about the way in which the mind made sense of those relationships. Implicitly, at least, it also suggested something about what the proper philosophical response to the fragmentation of knowledge ought to be. This was a plea for the importance of holding on to the big picture in science. Though he said little about them in the essay, the original lectures applied correlation to geology and physiology too. The theory of correlation implied that specialists needed a generalist to make sense of their experiments and observations. Grove was offering his audience at the London Institution, and the readers of the essay, an ambitious philosophical programme that emphasised the importance of remembering the big picture. Was he perhaps imagining a role for himself in that respect?

Being Useful

One of the main themes in Grove's inaugural lecture at the London Institution was the usefulness of science. That theme was still very much present in his theory of correlation too. Many of Grove's readers would have been familiar with the sorts of popular books on political economy that flourished during the 1830s and 1840s. Many of them would also have noticed the similarities between what Grove had to say in *Correlation* about the natural economy and what authors like Charles Babbage or Andrew Ure had to say about political economy (what we would now call economics). Books like these were widely read – not surprising during a period of such profound social and economic transformation. They formed part of a tradition going back to Adam Smith's *Wealth of Nations* that paid particular attention to the division of labour as the key to increasing economic productivity. The idea was that by dividing industrial processes into a number of separate tasks (pin making was Adam Smith's favourite example), the rate of production could be significantly increased whilst costs were reduced. Bigger factories made for more division of labour and relatively unskilled workers could be employed to carry out these relatively simple and repetitive tasks.

Authors like Charles Babbage in his *Economy of Machinery and Manufactures* emphasised the importance of new machines as much as the division of labour. Babbage, for example, emphasised the way in which the introduction of machinery provided factory managers with a new way of controlling the pace at which workers performed their allotted tasks. 'One of the most singular advantages we derive from machinery is the check which it affords against the inattention, the idleness, or the knavery, of human agents,' he said.[36] The electrician and mathematician Peter Barlow, inventor of a popular electromagnetic demonstration device called Barlow's Wheel, had something similar in mind when he argued that 'expense ... is not ... all that enters into the question of what is to be understood by the term economy. We have to look at the convenience of application, the steadiness and uniformity of power and motion.'[37] Writers like this often envisaged the factory as a kind of self-regulating machine made up of human and non-human parts. Barlow defined a factory as a place where 'a number of mechanical processes, depending entirely on their guidance on Machinery, and that Machinery driven by steam ... although, perhaps, not a hand is employed in making the article, but simply in feeding the fire-place and attending the Engine or Machines'.[38] Factories were places where machines put the powers of nature to work making goods and humans tended them.

Andrew Ure made exactly the same point in his *Philosophy of Manufactures*. 'Manufacture,' he said, 'is a word, which, in the vicissitude of language, has come to signify the reverse of its intrinsic meaning, for it now denotes every extensive product of art, which is made by machinery, with little or no aid of the human hand; so that the most perfect manufacture is one that dispenses entirely with manual labour.'[39] Ure's ideal factory was a huge, self-regulating machine. It was

> the combined operation of many orders of work-people, adult and young, in tending with assiduous skill a system of productive machines continually impelled by a central power ... [it] involves the idea of a vast automaton, composed of various mechanical and intellectual organs, acting in uninterrupted concert for the production of a common object, all of them being subordinated to a self-regulating moving force.[40]

Ure, then a lecturer at Glasgow's Anderson's Institution, had first made a name for himself in 1818 when he carried out electrical experiments on the body of an executed criminal, making his body convulse and his face grimace like 'the wildest representations of a Fuseli or a Kean'. Ure was obviously good at finding ways of making bodies act like machines.

Alert readers would have noted the similarity between Grove's account of a nature working through correlation and the picture of a factory as a self-contained system of production and exchange. His concern for the economics of electricity was evident in a lecture on the 'progress made in the application of electricity as a motive power' at the Royal Institution in February 1844. One of his aims in that lecture was to persuade his audience of just how economic was his nitric acid battery, despite the apparent expense of some of its components. He offered them some calculations to estimate the battery's efficiency:

Chemical equivalent of zinc, 32
Chemical equivalent of nitric acid, 54
54/3 = 18
32:18 :: 45:25.3
45 lbs. of zinc at 3d. = 11s.3d.
25.3 lbs. of real (i.e. 50.6 of commercial) nitric acid at 6d. = £1.5s.7½d.
11s.3d. + £1.5s.7½d. = £1.16s.10½d., the expense of one horsepower for twenty-four hours.[41]

He left the cost of the sulphuric acid out of the equation, since it was more than balanced by the value of the salts formed during the process. This was his battery reduced to its economics. He acknowledged that it remained far more expensive than steam, but what could one expect, he asked, 'as with one we use for fuel manufactured materials in the production of which coals, labour &c., have been expended; in the other coal and water are used directly'.

This was putting an economic value on correlation. The forces exchanged for each other by Grove in his London Institution laboratory using his battery had exchange values not just with reference to each other, but with the products of Victorian industry too. Natural philosophers in their laboratories in this picture were doing something very similar to what Victorian factory managers did. Theirs was the task of coordinating the transformation of forces that took place through their instruments and to make those transformations as efficient as nature allowed, just as it was the factory manager's job to make the process of production as efficient as possible. This was a view of natural philosophy that was steeped in industrial culture – not surprising for a Swansea man who had grown up in just such an environment. It was theoretical and practical at the same time. Grove was showing himself to be fluent in the language of abstract philosophy and the hard-nosed calculations of the industrialist at the same time. Thinking about causes, thinking about what went on in the laboratory and thinking about the machines of industry all turned out to be the same thing from the perspective of the correlation of physical forces.

Receptions

The *British and Foreign Medical Review*'s reviewer commended 'this outline of Mr. Grove's views' as

> one which is most fertile in topics of interesting speculation; whilst it shows how closely such speculations may be connected, by a logical mind, with phenomena obvious to our senses. We commend its study, to those who are capable of mastering it, as an example of the mode in which we trust that the phenomena of vitality may be ultimately generalized; when the same preparation shall have been made, by patient and accurate research, for the development of analogous views.[42]

The book went through six editions during Grove's lifetime, the second edition appearing only a few years after the first in 1850. The second edition contained significant additional material, particularly more experimental illustrations to add flesh to the bones of correlation. The *British and Foreign Medico-chirurgical Review* thought that the new edition was 'much more intelligible to those who are not prepared, by previously acquired knowledge, to enter into the author's views'.[43] The nonconformist *British Quarterly Review* gave the new edition a rave review, enthusing how the physical forces, as they are now usually denominated, are regarded as the active agents of creation; and notwithstanding the dissimilarity in their effects, they are, by many philosophers, regarded as modifications of some great primary cause or principle. The author of the 'Correlation of Physical Forces' has placed this hypothesis in the strongest light; he has given it the powerful support of his correct knowledge over an extensive field of inductive research, and the advocacy of an earnest and able pen.[44]

Though it seems clear that he discussed the matter in his original lectures there was very little in the first edition of *Correlation* about the implications of Grove's theory for understanding the relationship between physical and vital forces, beyond noting that 'the same principles and mode of reasoning as I have adopted in this essay, might be applied to the organic, as well as the inorganic world'.[45] The up and coming physiologist William Benjamin Carpenter quickly saw the implications nevertheless. His paper on 'the mutual relations of the vital and physical forces' was read to the Royal Society in 1850 and published in the *Philosophical Transactions*.[46] Carpenter drew heavily on Grove's work to make the point that the forces animating living bodies could be understood in terms of correlation too. This paper was, according to the *British and Foreign Medico-chirurgical Review*, 'the first *systematic* attempt that has been made, in this country at least, to work out the subject'.[47] Other authors such as Thomas Laycock and Alexander Bain took the idea further still, arguing that mental activity could be brought under the doctrine of correlation as well. In fact Laycock wanted to use the theory of correlation as applied to mind as the basis of a whole new system of morality.[48]

Correlation found its way into contemporary literature too. George Eliot drew on Grove's ideas when she was writing *Middlemarch*. Pondering the unexpected consequences of human actions, she suggested that 'in watching effects, if only of an electric battery, it is often necessary to change our place and examine a particular mixture of group at some distance from the point where the movement we are interested in was set up'.[49] Like Carpenter she was taking an imaginative leap from Grove's discussion of the interplay of inanimate forces and applying them to the business of life – in this case the complex interactions of early nineteenth-century society. More literally, Charles Dickens's friend, the popular and prolific novelist Edward Bulwer-Lytton, drew on Grove's views on correlation, along with Michael Faraday's discussion of the conversion of force, to describe the futuristic science of that mysterious race of beings, the vril-ya in *The Coming Race*. The vril-ya's superior science was due to the fact that they had 'arrived at the unity in natural energetic agencies, which has been conjectured by many philosophers above ground, and which Faraday thus intimates under the more cautious term of correlation'.[50] Throwaway remarks like these, like Charles Kingsley's allusion in *The Water Babies*, show just how widespread Grove's view on correlation had become during the second half of the nineteenth century.[51]

Another unexpected enthusiast for correlation was Karl Marx. 'I have in my hands the *Correlation of Physical Forces* of Grove. He is certainly the most philosophical of all English scientists (and German too!)' Marx wrote enthusiastically to Friedrich Engels in August 1864.[52] Engels drew on Grove in putting together the *Dialectics of Nature* too. The language of correlation remained pervasive throughout the second half of the century, to the dismay of some proponents of the new theory of the conservation of energy (remember Peter Guthrie Tait's dismissal of Grove's ideas as 'humbug'!). When James Clerk Maxwell reviewed the sixth edition of *Correlation* in *Nature* in 1874 he acknowledged the continuing power of Grove's concept. The book had 'a value peculiar to itself', he said. It would 'always retain its place in the memory of the student of human thought, as one of the documents that serve for the construction of the history of science'.[53] With a bit more acid, he suggested that Grove's theory had 'certainly exercised a very considerable effect in moulding the mass of what is called scientific opinion, that is to say the influence which determines what a scientific man shall say when he has to make a statement about a science which he does not understand'.[54] In 1846, however, this was all firmly in the future. Grove had resigned his plum position as professor of experimental philosophy at the London Institution, ostensibly due to the pressures of a growing legal practice. He was looking for other ways of making his way and his mark in the scientific world.

4

SCIENTIFIC REFORM

When William Robert Grove was elected a fellow of the Royal Society on 26 November 1840 he was joining one of Europe's oldest and best established societies for men of science. The Royal Society of London for Improving Natural Knowledge was founded on 28 November 1660 when a group of natural philosophers returned to London following the restoration of the monarchy and started holding regular meetings at Gresham College and decided to organise themselves into a 'Colledge for the Promoting of Physico Mathematicall Experimentall Learning'.[1] A royal charter soon followed, guaranteeing patronage, though very little cash, as things turned out, from the newly returned Charles II. The Royal Society was a society of gentleman virtuosos who modelled their new institution on Francis Bacon's ideal of a reformed and useful natural philosophy that could be placed at the service of the restoration state. Its founder members included Robert Boyle and Christopher Wren. The first curator of experiments, who had the task of demonstrating the latest experimental discoveries to the fellows, was Robert Hooke. Another early fellow was the diarist Samuel Pepys, who was elected in 1665. He became the society's president in 1684. It was during his presidency that the society took on the task of publishing Isaac Newton's *Philosophiae Naturalis Principia Mathematica*.

Newton had been elected a fellow of the Royal Society in 1672 and became its president from 1703 until his death in 1727. Under his presidency the society rose to new heights of prestige and influence. It was not just a society for men of science, but for gentlemen of means as well. Newton's successor as president, Hans Sloane, was a wealthy and fashionable physician who counted successive monarchs amongst his clients. A prodigious collector of antiquities, the collection he bequeathed to the nation on his death in 1753 became the foundation of the British Museum. Sloane was a master at the game of patronage, having amassed a string of prestigious appointments for himself in the course of his career, and in turn he used his presidency of the Royal Society to dole out patronage and return favours for himself. Being a fellow of the Royal Society was a mark of prestige, not just for men like Joseph Priestley or Benjamin Franklin who were active experimenters, but for society gentlemen as well. In 1780 the Royal Society moved from its quarters in Crane Court to apartments in Somerset House on the Strand. Housed in the heart of fashionable London it was to all intents and purposes a gentlemen's club for those with an interest in natural philosophy.[2]

One of the main architects of the society's move to fashionable Somerset House had been Sir Joseph Banks, recently elevated to the presidency. Banks came from a wealthy landed family and moved easily through the upper echelons of metropolitan society. He had been elected to a fellowship in 1766 and had made something of a name for himself as the naturalist on *HMS Niger*'s voyage to Greenland and Labrador. In 1768 he was appointed the naturalist on *HMS Endeavour*'s voyage of exploration to the south seas, captained by James Cook. The voyage made Banks famous. Following Sir John Pringle's resignation as president of the Royal Society in 1778, Banks, with his combination of wealth, influence and natural philosophical reputation, was an obvious candidate to replace him. He was to remain as president for an unprecedented forty-two years until his death in 1820. Banks certainly governed the Royal Society with an iron grip throughout his presidency. What state patronage there was for science during this period was his to dispense. His own earlier voyages had been funded by the admiralty – the main purpose of the voyage with Cook on the *Endeavour* had been to observe the transit of Venus from the island of Tahiti in 1769, for example, as well as explore new territories with an eye to future economic exploitation.[3]

Banks was adept at making full use of his networks of power and influence. He was not just the president of the Royal Society, but also wielded considerable influence over the affairs of the Antiquarian

Society, the Horticultural Society, the Linnean Society and the Royal Institution. Beyond that spectrum of intellectual groups, he was a key player in the business of the Board of Agriculture, Kew Gardens, the British Museum and the Board of Longitude. He had the ear of the royal court as well. Banks's success as president rested in large part on his ability to get all these different groups and organisation pulling in the same direction. He had made himself the go to man of late eighteenth-century and regency metropolitan science. Without his support it was very difficult to get anything done. Towards the end of his reign though, Banks's authoritarian dominance led to increasing resentment on the part of a new generation of up and coming men of science. His firm grip on the purse strings for science made it very difficult to find support for any initiatives that failed to gain his approval. One result of this was that frustrated opponents started looking for opportunities to establish their own rival institutions. The geologists declared their independence in 1807, founding the Geological Society. The astronomers followed suit in 1820 with the Astronomical Society.[4]

Banks's death on 19 June 1820 offered the growing band of dissidents their opportunity to wrest control of the Royal Society from the hands of the *ancien régime*. The natural philosopher and chemist William Hyde Wollaston acted as caretaker for the rest of that year before Sir Humphry Davy was elected as president. Davy was meant to be a compromise candidate between the old guard and the reformers.[5] He had close links with Banks and his cronies on the one hand and on the other had the sort of scientific credentials that should satisfy the new generation of reformers, who resented the dominance of men they regarded as scientific dilettantes. He had a foot in both camps and as a result he satisfied nobody. Astronomers were particularly incensed over the Royal Society's dominance over the Board of Longitude, which in turn controlled state patronage over a range of astronomical activities. Men like Charles Babbage and John Herschel – son of Uranus's discoverer – had their own ideas about how astronomy should be done and wanted access to the Board of Longitude's resources. Under Davy's presidency some of the leading reformers, including Babbage and Herschel, gained a place on the Royal Society's council. They were starting to develop their own power base.

Under the Banner of Science

In May 1827 the Royal Society established a committee 'to consider the best means of limiting the number of members admitted into the Society, and to make such suggestions as may seem to them conducive to the welfare of the Society'.[6] This may sound innocuous enough, but it was no coincidence that the committee was packed with reform-minded members of the Royal Society's council. Their goal, or more accurately their chosen means to achieve their eventual aim was limitation, as they called it. The traditional process for the election of new fellows was well established and straightforward. Potential fellows were nominated by the existing fellowship, their names and the list of their supporters was posted at Somerset House. The list of names of nominees was then read out at the next meeting of the society, and if they gained the support of two-thirds of the fellows present and voting (there were rarely any objections) they were elected. There was no limit to the number of fellows that could be elected in any year, and no limit to the size of the fellowship as a whole. The reformers wanted to do away with all this. They wanted a limit to the size of the fellowship and a limit to the number of fellows that could be elected annually.

The report the committee presented to the council a little over a month later on 25 June outlined a radical series of proposals for reforming the Royal Society's arrangements. The central recommendation was that the number of ordinary fellows should be limited to 400. To achieve this, no more than four new fellows would be elected each year until the target reduction was reached. According to the committee's report, the new arrangements would make sure that the fellowship's composition would be a 'fair representation of the talent of the country, the consequence of which will be, that every vacancy would become an object of competition among persons of acknowledged merit'.[7] Furthermore, the committee recommended that the election of new fellows should take place only once a year by means of an alphabetical list of candidates that would be drawn up by the council. The list would be voted on by the fellows present and the four candidates that received the most votes would then be presented to the meeting again for a second vote to ensure that each of them received the two-thirds majority required by the Royal Society's charter. Almost as an afterthought, the committee also recommended that from then on the composition of successive councils should be a matter of discussion by the expiring council rather than being the prerogative of the president as had traditionally been the case.

If all this had worked out, it would have been a very clever coup. The proposals if accepted by the council would have ensured reformist dominance for the foreseeable future. The stringent limitation on new fellows would have deprived the old guard of their opportunity to gain favour and patronage by proposing the politically powerful and wealthy as fellows of the Royal Society. The provision that the council, rather than the president, should choose their successors meant that the hoped for reformist dominance could be maintained indefinitely, regardless of who was president. Unfortunately for the reformers, however, things did not quite work out as planned. In a clever delaying tactic, opponents persuaded the sitting council to defer any decision and leave everything to 'the most serious and early consideration of the Council for the ensuing year'.[8] By the end of 1827 the increasingly beleaguered and ill Sir Humphry had resigned the presidency – he would be dead two years later. After much skulduggery Davy's fellow Cornishman Davies Gilbert was elected as his successor – and he had been one of the two dissenting members of the committee behind the reform proposals. He in turn made quite sure that the composition of the new council was favourable to his own anti-reform views. The *ancient régime* triumphed and the committee's report was kicked into the long grass indefinitely.

The reformers had been thoroughly outmanoeuvred by a group of men who were, after all, extremely skilled at playing the game of power. With Gilbert as president there was little opportunity for further attempts at routing the old guard. That opportunity came in 1830 when Davies Gilbert relinquished the presidency. He and his allies wanted him replaced by the duke of Sussex, younger brother of the king. It would be the utmost act of patronage on the part of the Royal Society's traditionalists – an assertion of their dominance over the society that would give them a powerful voice right at the heart of the royal court.

The opposition were predictably scandalized by this raw assertion of patronage. John Herschel, son of the eminent astronomer, was put up as an alternative candidate for the presidency. It was a hard and often vicious campaign and in the end the traditional camp triumphed, the royal duke of Sussex became president of the Royal Society.[9] Charles Babbage's response was to pen his *Reflections on the Decline of Science in England*. It was a vituperative attack on the Royal Society leadership, accusing them of corruption and nepotism on an industrial scale. They were men more interested in feathering their own nests than in furthering the interests either of the society they represented or of science as a whole.

It is no coincidence, of course, that this battle for the soul of the Royal Society took place when it did. Throughout the 1820s calls for reform resonated across the cultural and political landscape of Britain. Parliamentary reform might have taken centre stage as radicals and reformers fought for what would eventually become the Great Reform Act of 1832, but that campaign shared the same language as campaigns to reform other moribund institutions like the Royal Colleges of Physicians and of Surgeons, as well as the Royal Society. Thomas Wakley, the firebrand editor of *The Lancet*, called the leaders of the Royal Colleges 'crafty, intriguing, corrupt, avaricious, cowardly, plundering, rapacious, soul-betraying, dirty-minded BATS', using language that might have made even Charles Babbage blush.[10] Babbage certainly pulled few punches in his *Reflections*. It was a thoroughgoing denunciation of the Royal Society's leadership at all levels. Babbage and Herschel, in their younger days at least, had decided republican sympathies. Herschel declared himself a 'poor snivelling democratic dog' in the aftermath of Waterloo.[11] Babbage's vision of the Royal Society transformed into a state-run and state-funded academy and its fellows as salaried civil servants betrayed his Bonapartist sympathies too. But just as the campaign for parliamentary reform lost momentum after its partial victory in 1832, the campaign for reform of the Royal Society fell apart after its defeat in 1830 – at least until the 1840s when the banner of scientific reform was unfurled again by a new generation, and with Grove taking on a leading role in the battle.

Its Present Form is Not Wholesome

It was not long after his election as a fellow of the Royal Society that Grove started to express some misgivings about the direction and purpose of the august organisation of which he had become a member. In 1841, as noted a few chapters ago, Grove was furious that one of his papers had been rejected for publication in the Royal Society's *Philosophical Transactions*. The angry letter he wrote to Michael Faraday made it clear that what particularly offended him was the realisation that he would need to 'make interest' to gain publication.[12] In other words, that the merit of his work in itself was not enough – he would have to curry favour to get what he wanted. Grove, quite clearly, did not want to play the Royal Society's game of patronage. It cannot have helped his growing disenchantment with the society, when, a year or so later, he found himself involved in an acrimonious exchange of letters in the pages of the *Philosophical Magazine* with John Frederic Daniell, professor of chemistry at King's College London and the Royal Society's foreign secretary. Daniell had alleged, in the context of a priority dispute with the French natural philosopher Edmond Becquerel regarding the invention of the constant voltaic battery, that 'Prof. Grove has never spoken of his battery but as the further application of principles which I had previously deduced'.[13]

Grove's riposte to the allegation was revealing. He may have said something along those lines in lectures at the London Institution, he conceded, but only because he was flattered at such an eminent man's presence there and wanted to flatter him in return. He had, perhaps, paid 'a greater compliment than the occasion required' and had gone out of his way to avoid mentioning any disagreement, thinking that to do anything else would have been 'singularly bad taste'.[14] Grove was unhappy that his name had been dragged into a dispute that was nothing to do with him. He remarked ironically that 'Professor Daniell has (probably for good reason) avoided my name when making copious use of my results; I wish he had carried out this principle and not introduced my name into a controversy with which I have nothing at all to do.'[15] Grove was clearly using Michael Faraday as a sounding board for some of his unhappiness with the way the Royal Society and some of its officers were treating him. Faraday was sympathetic. 'As to the Royal Society,' he replied,

> you know my feeling towards it is for what it has been and I hope may be. Its present state is not wholesome. You are aware that I am not on the Council and have not been for years and have been to no meeting there for years. I do hope for better times.[16]

In an anonymous contribution to *Blackwood's Magazine* a few months later, Grove had his opportunity to let rip about what he thought of the state of metropolitan science and its institutions. Yes, he said, scientific societies could be good things

> insofar as they carry out their professed objects of facilitating intercourse between the votaries of similar branches of study – they do good by the more attainable communication of the researches of those who cannot afford, or will not dare, the ordinary channels of publication ... they give an *espirit de corps*, which forms a bond of union to each section, and induces a moral discipline in its ranks.

But, by the same token, they

> do harm by the cliquery they generate, collecting little knots of little men, no individual of whom can stand his own ground, but a group of whom, by leaning hard together, can, and do, exercise a most pernicious influence; seeking petty gain and class celebrity, they exert their joint stock brains to convert science into pounds, shillings and pence; and when they have managed to poke one foot upon the ladder of notoriety, use the other to lick furiously at the poor aspirants who attempt to follow them.[17]

It was a wholesale condemnation not just of the Royal Society but of the whole institutional structure of English science.

Figure 3 Somerset House, Meeting of the Royal Society, engraved by H. Melville after a picture by Fairholt, published in *London Interiors*, 1845.

What was needed, Grove said, was an institution modelled along the lines of the French Académie des Sciences, but, unlike that organisation, immune from overt state and political control. Grove was well aware of the suspicion with which many of his fellow men of science regarded the prospect of too close a connection between scientific bodies and the state. He regarded the politicking of some of his French counterparts with more than a little suspicion himself. That was not one of the aspects of French scientific culture he wanted to emulate. 'Politics are already too much mixed up with all government appointments in England' already, he thought, 'their influence is at present scarcely felt in science, and we would not willingly risk an introduction so fraught with danger'. What he wanted was the remodelling of the Royal Society into an institution 'accessible only to men of high distinction who would be thus constituted the oligarchs of science'. The reformed society would be a state institution, organised and financed by the state, but independent of it. It would therefore be immune to any mere 'political excitement'.[18] This was Grove's blueprint for the reform of science: a powerful centralising body that could direct science as it wished, governed by an acknowledged elite of scientific oligarchs.

In 1845 Grove had an opportunity to realise this fantasy. He was appointed a member of the Royal Society's council. Within a year, some of his fellow council members were making their own moves to try again to reform the Royal Society's arrangements for electing new fellows. In some ways, this was to be an attempt to resurrect the buried report produced back in 1827, though there were important differences as well – and some supporters of reform back in the 1820s were opposed to it now. It seems not implausible, at least, that one of the reasons underlying Grove's own selection as a council member was his support for reform and that it was a move on the part of the reformist camp to bring a useful fellow traveller on board. Using the Royal Society's council minutes, the minutes of the charter committee and Grove's own correspondence with his co-conspirators it is possible to reconstruct in some detail the convoluted backroom wheeling and dealing that eventually led to the implementation of the reform proposals. It is worth doing in this instance because it casts important light on Grove's place in the networks of influence that governed London science during the 1840s. As it turns out, these two years or so of back and forth arguing over reform were to be the make or break years of Grove's own scientific career too.

At the council meeting of 7 May 1846, Leonard Horner proposed two resolutions to the effect that 'it is expedient to revise the Charters of the Royal Society, with a view to obtaining a Supplementary Charter from the Crown' and that 'the President, Vice-Presidents, Secretaries, and Foreign Secretary be appointed a Committee to consider and report what alterations it would be desirable to introduce into the Supplementary Charter'.[19] Horner was a veteran of the campaign to elect Herschel as president in 1830, and in 1846 he was the president of the Geological Society. He was politically well connected (his brother, Francis Horner, had been a prominent Whig politician) and had been a member of several parliamentary commissions investigating matters such as sanitation and factory legislation. He had also been the first warden of London University – that 'godless institution in Gower Street'.[20] In a letter to Peter Mark Roget, who was then the Royal Society's senior secretary, he made it clear that he felt that in the interests of efficient administration the society's charter needed a thorough overhaul and he also made quite explicit his belief that what he viewed as the inflated membership of the society had to be cut to maintain the society's position as an exclusive body.[21]

Other members of the charter committee saw things differently. At its first meeting on 11 May, only four members were present: the marquis of Northampton in the chair and George Rennie, Peter Mark Roget and Edward Sabine. Six 'principal objects' were identified by the committee members present. First, they suggested that the day prescribed by the charter for holding the annual election of the president, officers and council of the Royal Society should be changed from St Andrew's Day to some more suitable date in the months of April, May or June. This was presumably at the suggestion of the marquis of Northampton who felt that the St Andrew's Day Anniversary was inconvenient from the point of view of a man who habitually moved in court and parliamentary circles. The committee also recommended that a specific quorum of fellows present should be required to constitute both ordinary meetings and meetings of council; that ordinary members of council should be able to chair meetings in the absence of the president or vice-presidents; that council should be empowered to fill by ballot any vacancies that might occur amongst its members during the term of office and that the council should be given the power 'to apply at their discretion the funds of the Society in furtherance of the objects of the Society'. Finally it was recommended that 'no person shall be elected a Fellow unless he has in his favour 5/6th of the members voting'. The previous requirement had been that two-thirds of the votes cast should be in favour.[22] Compared to some of the reformers' ambitions, this was trivial stuff.

Leonard Horner was firmly adamant that what was required was a strict limitation on the admission of new fellows and told Grove so in no uncertain terms:

> I do not believe that anything short of a limitation of the numbers will place the royal Society in that high degree of estimation in which it ought to be held both at home and abroad. Neither in your letter nor in one I had from Dr. Roget does there appear any reason to have been assigned for not assenting to the limitation except the apprehension of financial difficulties, which my calculation seemed to show there is no ground for, and it does not appear that I have made any material mistake.[23]

Only a few days later he suggested that Grove should propose a motion favouring limitation at the next council meeting of the Royal Society. When the next meeting of council duly took place on 28 May, three crucial resolutions were passed. The council instructed the charter committee to examine in detail the current state of the charter and statutes, and report upon the extent to which they were suitable to the current state of the Royal Society. They were instructed to consider whether any alteration would be best made by alterations to the statutes or to the charter. Grove was appointed a member of the charter committee and it was proposed (presumably by Grove) and resolved that at the next meeting of council 'it be considered whether in the event of the Society obtaining a new Charter, it would be expedient to introduce a Clause limiting the number of Fellows to be elected in any one year, and another Clause to alter the mode of election'.[24] In other words, the council threw out the committee's recommendations and told them to start again.

The minutes give little indication of what was discussed at the next meeting, noting only that C. R. Weld, the Royal Society's assistant secretary had been instructed to prepare an abstract of the relevant provisions of the charter and statutes, and that instructions concerning the committee's brief had been duly received from the council. A similar instruction 'to consider whether any, and what improvements can be made in the mode of electing Fellows' was repeated to the committee after the next council meeting on 4 June.[25] Matters proceeded rapidly at the next few meetings of the charter committee. The report of the 1827 committee was discussed and Horner was instructed to prepare a report to be presented to the full council. The brief introductory section of this report made no secret of the fact that its proponents regarded the limitation of new fellows as essential. The report made it clear that the instruction that they should 'consider whether any and what improvements can be made in the mode of electing fellows' had been regarded as the most important:

> Your Committee directed their attention in the first instance to the last received instruction ... being satisfied that there is no subject which at present more nearly concerns the best interests of the Society. It appears from the Minute-Books of the Council, that on various occasions the expediency of limiting the number of Fellows has been under consideration and strongly urged, and your Committee having carefully deliberated upon that important question have come to the conclusions which we now beg leave to submit to the consideration of the Council.[26]

Of the fourteen recommendations, ten were solely concerned with the question of limitation. The first two recommendations simply stated that 'the election of ordinary Fellows not included in the privileged classes referred to in Section 4 of Chapter I of the Statutes, shall take place on one day only in each year, and on the third Thursday of June', and that 'the number to be elected in any one year shall not exceed fifteen, exclusive of the privileged classes to be referred to hereafter'. The bulk of the report then dealt in detail with the proposed means of administering this new mode of election. The committee suggested that it should be possible to implement these 'most important and desirable' changes by modifications to the statutes without recourse to a new or supplementary charter, though they acknowledged that a proper legal opinion should be obtained on this point and that some modifications to the plan put forward might be expedient in order to avoid the trouble and expense of obtaining alterations to the charter. The report emphasised, however, 'that the interests of the Society require that measures be forthwith adopted, either by Charter or Statute, for checking the too great increase of Fellows by the election of persons who may not be qualified to fulfil the objects of the institution of the Royal Society'.[27] Further discussion of the matter was deferred until after the society's summer recess and the pro-limitationist camp took full advantage of the opportunity for extensive lobbying.

Despite his presence at the first abortive meeting, Edward Sabine, the Royal Society's foreign secretary, had now joined Horner and Grove in pushing for a strict limitation on the admission of new fellows. He was certainly aiding their attempts to secure the support of as many fellows as possible for their reform proposals. In many respects Sabine was a rather surprising ally – he had after all been publicly accused of fraud by Babbage in his 1830 *Reflections*. That Sabine was now on board offers one hint, at least, that Grove and his fellow reformers were looking for something rather different from the reforms espoused by the previous campaigners. In any case, the lobbying efforts of Grove, Horner and Sabine met with some success. On 28 October, just a few days before the Royal Society was due to reconvene after the summer recess, Horner wrote to Grove enclosing a letter from Sabine reporting that all opposition to the charter committee's report had been abandoned.[28] As it turned out, Horner's and Sabine's optimism was a little premature. At the council meeting on 5 November there was a concerted attempt to out-manoeuver the pro-limitation lobby by proposing that the promised discussion of the charter committee's report should be indefinitely adjourned. The reformers were successful in seeing off this attempt at consigning their report to the same fate as the 1827 proposal, and it was resolved instead: 'That the Council approves of the two recommendations in the report of the Charter Committee; viz. that the number of Fellows to be elected in any one year do not exceed fifteen, and that the Council do recommend to the Fellows the most eligible Candidates.'[29] The council also reaffirmed its previous resolution that the charter committee should reconvene in order to prepare a case for counsel to ascertain whether or not the projected changes could be achieved without a change of charter.

It was becoming clear what, or rather who, lay behind the attempted sabotage. Peter Mark Roget was using his position as secretary to delay the proceedings for as long as possible. Horner described his behaviour as 'unworthy of one occupying the position he does in the Society'.[30] As the senior secretary Roget was responsible for the minutes of council meetings, and Horner was incensed when he discovered on receiving a draft of the unconfirmed minutes of the last council meeting that they were incomplete: 'The motion of Col. Sabine for the appointment of a Committee is not inserted' was his complaint. Grove and he should call on Roget urgently to rectify the matter, he suggested.[31] When the committee was appointed in mid-1846, Roget had already been secretary of the Royal Society for over eighteen years, having been first elected on 30 November 1827. As an officer of the society he had been publicly vilified by Babbage, who had accused him of falsifying Royal Society council minutes.[32] His experiences were unlikely to have left him particularly sympathetic to those who marched under the banner of constitutional reform. On the contrary, as the senior officer, in terms of years, of the Royal Society his intimate knowledge of its machinery and his long apprenticeship in the task of manipulating its well-worn levers were hardly calculated to infect him with any great enthusiasm for changing the gearwheels.

Eventually, however, Roget was pressurised into sanctioning a meeting of the charter committee to be held on 16 November but the consultation with the Crown lawyers did not take place until 24 November. At this meeting the pro-limitation group received a further blow. The Attorney and Solicitor General gave their opinion as follows:

> The point submitted is not free from doubt; but, after the best consideration we can give it, we are of opinion that the Council cannot, by virtue of the general power of regulating the body given to them by the Charters, pass a Statute limiting the number of ordinary Fellows to be elected in any one year.[33]

Swift action would need to be taken to overcome this latest – and potentially disastrous – hurdle, particularly since a meeting of the charter committee was to be held the following day, followed on 26 November by the last council meeting before the annual anniversary meeting. On the very day of the consultation therefore (according to correspondence from Horner to Grove) Grove was already circulating a report to both Sabine and Horner containing an outline plan to institute slightly altered revisions to the statutes so that limitation could be effectively achieved without the necessity of changing the charter.[34]

Grove's role at this stage must have been crucial. As a practising lawyer he was unquestionably in a better position than anyone else on the committee to come up with an alternative set of recommendations. It is therefore presumably no coincidence that the fragment of the committee's draft report inserted with the charter committee minutes is in Grove's handwriting, since he was almost certainly responsible for much of its content.[35] The solution Grove suggested was ingenious: rather than imposing a strict limitation of fifteen new fellows, the council should instead simply 'recommend to the Fellows the most eligible Candidates; such selected Candidates not to exceed fifteen in any one year'. The presumption being that the likelihood of the fellows present at the meeting agreeing on the same non-listed candidates in sufficient numbers to ensure their election was so small as to be discounted.[36] Time was now of the essence. 'Delay must be strenuously resisted,' Horner warned, 'and we must have it settled that the meetings of the Council shall be held so as to make it possible to have the General Meeting early in February.'[37] It was finally agreed that 'the Council be specially summoned on the 14th of January to consider the recommendations contained in the Report of the Charter Committee'. This meeting proved to be the decisive engagement of the campaign.[38]

At this special meeting of the council, Horner presented the charter committee's fourteen recommendations as a series of resolutions, but there was so much fuss over the proposal to limit the number of new fellows that it was the only resolution discussed. First of all it was proposed (probably by the president, the marquis of Northampton) that the resolution omit the words 'not exceeding fifteen'. When this amendment was defeated, it was proposed that the word 'fifteen' in the resolution should be replaced by 'twenty'. This was defeated too.[39] They reconvened to consider the remaining resolutions a week later on 21 January. With the sticking-point of limitation out of the way, there was little left to fight over. All the remaining resolutions were passed with no opposition and the officers of the society, along with Grove and Horner, were appointed to draw up the new statutes.[40] These were presented to the council for their approval on 28 January and less than two weeks later on 10 February 1847 the relevant sections of the old statutes were repealed and replaced by the new proposals. It was not at all plain sailing, however: Horner's correspondence with Grove remained full of complaints about Roget's behaviour. 'To what lengths will he not go!' he speculated.[41]

The procedure set out by the 1847 statutes for the election of new fellows was that at the first ordinary meeting in March of any year, the names of all candidates nominated since the first March meeting of the previous year would be read out by the secretary in alphabetical order. Following this, in the first week of April a printed list of all the names read out, along with the names of their proposers, would be prepared and circulated to all ordinary fellows. From this list of candidates the council would then select by ballot the names of not more than fifteen candidates to be recommended to the society for election, the quorum for such a council meeting being eleven. This list of recommended candidates would then be read out at the first meeting of May and subsequently circulated to all the fellows along with a notice of the day and hour of election. This printed list of recommended candidates functioned as the ballot paper for the election of fellows. At the annual meeting all fellows present voted by returning this printed list of candidates to either the secretaries or one of the scrutators appointed at the meeting, having made any alterations to the list that they might see fit to make.[42]

Aftershocks

Even after the charter committee's recommendations concerning limitation had been fully incorporated into the Royal Society's statutes some opposition to the new measures continued to be expressed. The full extent of the marquis of Northampton's antipathy towards limitation was fully revealed at the following anniversary meeting on 30 November 1847. During his presidential address Northampton took the quite unprecedented step of publicly dissociating himself from the council's actions:

> During the last year, an important alteration has been made in our Statutes with reference to the election of new Fellows, as you must be well aware. This change was made with the approbation of a large majority of your Council. As I was one of those who entertained considerable doubts of its prudence and expediency, I cannot claim any praise should it prove advantageous to the Society, nor must I be considered responsible in case of failure.[43]

Northampton went on to remind his listeners that far more stringent measures had originally been contemplated and gave assurances that had they not been abandoned he would have made every effort to protect their privileges by insisting that the council should refer any final decision to a general meeting. He did not miss the opportunity to remind the fellowship that the council list of suitable candidates was a recommendation only and that they as fellows were perfectly entitled to undermine the new statutes by ignoring the council selection. It must have stuck in Northampton's craw that one of the official duties he was required to perform during the course of his address was awarding the society's Royal Medal for that year to no other than Grove himself.

Northampton repeated many of these points, including his 'own doubts on the main feature of these regulations' at the first June meeting for the election of new fellows according to the revised statutes in 1848. In what was to be his final address to the society, Northampton made no secret of the constituency which he felt he represented. He attacked those who proposed schemes for alterations in the society's structure and defended the role of polite learning within the Royal Society:

> As ... we have to support ourselves, any such extreme restriction of our members as should make us a very select body would deprive us of the power of publishing those Transactions, which constitute the main part of our scientific usefulness. Our body consists not only of men of science, but also to a certain extent of literary men,

eminent artists and gentlemen of rank and station. I believe that this widening of the basis of our Society is most useful to it, and besides serves the important purpose of enlisting in the cause of science those who may patronize and defend, though they do not follow her.[44]

His shot was well calculated to hit its target: the inefficiency of the *Philosophical Transactions* was a constant source of complaint from the reformist camp. Again he forcefully reminded the fellowship of their right to ignore the council recommendations but drew their attention also to the injustice that would occur 'if the supposed errors of these regulations were to be visited on the unoffending Candidates'.[45]

Some of the audience, however, did not share the finer sensibilities of the marquis of Northampton. After the statutes for the election of fellows had been read and scrutators appointed for the election, A. J. Stephens and John Lee promptly moved that the meeting should adjourn immediately. Since the new statutes do not seem to have made any provision for recalling the meeting for the election of new fellows on any day other than the third Thursday in June, this would effectively have prevented any new ordinary fellows from being elected during the course of that year. The motion was however defeated and the meeting proceeded dutifully to elect the list of candidates presented by the council.[46] A similar attempt was made at the next annual meeting for the election of fellows. On this occasion, William Tooke, seconded by John Lee proposed that

> the election of Fellows be adjourned until Thursday the 21st instant at three o'clock, and that it be recommended to the Council that the list for such election shall comprise the names of all the Candidates, designating those selected by the Council in such manner as may be deemed fit.

The effect of this motion had it been passed would have been to make it far easier for the fellows present at the meeting to reject those candidates selected by the council and replace them with others of their own choice. The motion was however defeated by an amendment proposed by Northampton, seconded by de la Beche, that 'the Society do now proceed to the election of Fellows'.[47] This appears to have been the last occasion at which explicit attempts were made to undermine the administration of the new statutes.

Grove and his co-conspirators had won the battle for reform. They had succeeded in seeing off strenuous opposition from the president and the senior secretary to get their own way. But it is undeniable that they had made some dangerous enemies along the way as well. Northampton and Roget remained powerful figures within the Royal Society – even though both were shortly to resign in response to their failure to defeat the reformers. A significant proportion of the fellowship was unhappy with the way things had gone too, including even some former supporters of reform campaigns. William Tooke, for example, had played a prominent role in John Herschel's presidential campaign back in 1830. This time, he was an opponent. For Grove though, the successful campaign must have seemed to open up new horizons. He had demonstrated his mettle as a political campaigner and forged valuable alliances with other powerful men within the metropolitan scientific elite. He was in a position to become a power behind the throne at the Royal Society and to nudge it further in the right (as he saw it) direction. In the meantime, he had other things to think about. His focus for the rest of that summer was not London, but Swansea. The British Association for the Advancement of Science was coming to town.

5

SWANSEA SCIENCE

When the British Association for the Advancement of Science (BAAS) arrived in Swansea at the beginning of August 1848 for its annual meeting it was seventeen years since its inaugural meeting in York in the summer of 1831. In those intervening years the association's gatherings had become an important part of the life of British science.[1] From fairly humble origins the British Association had boomed to become a huge affair. For a week in summer, natural philosophers and their entourages descended on a provincial town and turned the place into the temporary capital of British science. Locals and visitors flocked in their hundreds to scientific lectures delivering the latest developments in the various sciences, as well as to a packed itinerary of soirées and excursions. Local worthies had their opportunity to rub shoulders with the scientific lions. Meetings were extensively reported upon in the local and the national press. Hosting a gathering of the BAAS had become a mark of civic prestige and ambition. The association moved from town to town (though never to London) on an annual basis. Those towns competed for the honour of playing host to the assembled gentlemen of science. Inviting the scientific jamboree to Swansea was a deliberate signal of the town elite's ambitions and Grove, it seems, was the instigator.

The BAAS's first gathering in York was the result of hard work and perseverance by William Vernon Harcourt. Harcourt was a local clergyman with impeccable credentials. His father had been archbishop of York and his mother the daughter of the marquis of Stafford. After completing his studies at Oxford in 1811 he had entered the church and by 1824 he was a canon of York minster as well as the rector of a couple of Yorkshire parishes. In that year he was also elected a fellow of the Royal Society. There was more to Vernon Harcourt than the kind of scientific dilettante despised by the scientific reformers, nevertheless. In 1822 he had been one of the leading local lights behind the establishment of the Yorkshire Philosophical Society, and he was one of its first presidents. The Yorkshire society was typical of the kind of scientific organisations that were proliferating across the British Isles during the first half of the century.[2] In some ways, what Vernon Harcourt had in mind when he issued his invitation for natural philosophers to gather at York in 1831 was a means of bringing together these different societies. It certainly exceeded expectations. Several hundred people attended the meeting to experience the mixture of scientific lectures and entertainments that were to become a feature of the association's meetings.

The following year's meeting took place in Oxford. By now, it had become clear to the leaders of metropolitan science too that this was an organisation that could be moulded to serve their own interests. Charles Babbage had attended the York meeting and saw in the new association the possibility of providing a rival for the Royal Society. Babbage was already a fan of this kind of gathering. In 1828 he had attended a meeting in Berlin of the Society of German Naturalists and Natural Philosophers and thought that the fledgling British Association could provide the seed for a similar society in Britain.[3] The British Association could be turned into a powerful tool for scientific reform. It could provide an alternative view of how science in Britain might be organised and provide an alternative forum for those who were disenchanted with the Royal Society's corruption. From Oxford in 1832 the meeting moved on to Cambridge the following year. It was there, in response to a complaint from a curmudgeonly Samuel Taylor Coleridge that William Whewell coined the word 'scientist' to describe the attendees.[4] The BAAS's committees and leadership were dominated by the end of the decade by the good and the great of London science.

The annual meetings had by now assumed a regular and familiar routine. Visitors would know what to expect. As far as we know, Grove attended for the first time in 1839 when he presented an account of his new nitric acid battery to the chemistry section at the Birmingham meeting where, incidentally, Vernon Harcourt served as the association's president. Section B, the chemistry section, was presided over by Thomas Graham. Meetings were organised into several sections: section A was for the mathematical and physical sciences; section B was chemistry and mineralogy; section C was geology and practical geography; section D was zoology and botany; section E was medical science; section F was statistics; and section G was mechanical science. Attendees like Grove could offer papers in the various sections as well as rub shoulders with the associations' leaders at evening soirées and entertainments. There was something for everyone, and by 1837 the organisation had even achieved the dubious notoriety of being satirised by Charles Dickens as the Mudfog Association for the Advancement of Everything.[5] Dickens poked fun at the association's apparent eclecticism, but in reality the gentlemen of science made sure that everything was under careful control. It is tempting to wonder if Grove had started thinking as early as this, his first meeting, about the possibility of inviting the British Association to Swansea.

Such a move would certainly fit well with the clear ambitions that the Swansea circles in which he moved had for their town.[6] The plans that were already underway to transform the Swansea Literary and Scientific Institution into the far more ambitious Royal Institution of South Wales were an unambiguous pointer of the way that Grove and his peers were thinking. An invitation to the BAAS would make perfect sense from such a perspective. A visit to Swansea by the association would offer unrivalled opportunities for putting Swansea not just on the scientific, but the wider cultural map of Britain. It would show that Grove's birthplace was not just some obscure corner of an obscure country on the edges of empire, but a civilised and progressive metropolis. A visit by the BAAS would provide opportunities for forging and consolidating alliances of mutual benefit on a local and a national stage. Those running the show in Swansea would be able to reward their local supporters as well as display their credentials to the visitors. They could bring the scientific world to Swansea for a week in the summer and show the world that Swansea science was a force to be reckoned with.

Setting the Scene

The BAAS's arrival in town on 9 August 1848 was the culmination of several years of planning and negotiation. In 1846, the council minutes for the Southampton meeting of the association noted

> that a deputation has been appointed by the Mayor and Corporation of Swansea, the principal inhabitants, magistracy and country gentlemen of the neighbourhood, and by the Members of the Royal Institution of South Wales, to attend the Meeting at Southampton, for the purpose of inviting the British Association to hold their annual Meeting at Swansea at as early a period as may suit their convenience.[7]

The decision to accept the invitation was not a foregone conclusion. Swansea's was not the only offer on the table – the popular English spa town of Cheltenham was one of the competitors who had expressed an interest in hosting a meeting. On 12 September 1846, the *Cambrian* newspaper informed its readers that 'South Wales is to be honoured with a meeting of this learned body much sooner than could have been expected.' They reported that 'Memorials from the Corporation, neighbouring Gentlemen, Royal Institution, &c., most respectably and numerously signed, were forwarded to Southampton, for presentation by W. R. Grove, Esq., F. R. S., urging the claims of South Wales, and of Swansea in particular', to host a meeting of the BAAS as soon as practicable.[8]

A letter newly arrived from Grove explained that 'it is to go to Oxford next year; but I am happy to say, that we may confidently expect it to visit Swansea in 1848'. It was clear, though, that some council members thought that this little town on the edges of civilisation simply did not have the resources to host a gathering. The BAAS secretary, John Phillips, was instructed to visit 'for the purpose of examining and reporting on its means of public and private accommodation'. Phillips duly did so and reported back to the Oxford meeting the following year that the

> inquiry which you requested me to make, as to the accommodation which might be found in Swansea for a Meeting of the British Association has been rendered comparatively easy, and capable of an accurate answer, the excellent arrangement of Mr. Grove and the zealous assistance of his friends at Swansea

And that 'suitable accommodation for the public purposes of the Association could be found in the Royal Institution, the Town Hall, Theatre, and certain large school-rooms'. He noted that 'a Meeting of the Association at Swansea would have been quite impracticable, but for the prior establishment and creditable support of the Royal Institution of South Wales, which is there in operation'. He was impressed by the 'strength of united public feeling' which was 'at present undoubtedly strong, and in the right direction, and there is reason to believe it will remain so'.[9]

Swansea in 1848 was still recovering from a very turbulent few years. Just three years before Grove went to Southampton with the town's invitation to the BAAS in his pocket, Swansea and the surrounding district was embroiled in the Rebecca riots. Groups of men in skirts, dressed as the hosts of Rebecca, roamed the countryside at night, attacking and burning the hated turnpikes.[10] Grove, away most of the time in London, would have had little involvement with these troubles, but many of his Swansea friends were up to their necks in the business. Lewis Weston Dillwyn and John Dillwyn Llewelyn his son and Grove's near contemporary, were, as local magistrates, charged with leading the fight against the rioters. There were concerns that Llewelyn's palatial property at Penllergaer might be burned to the ground by the mob. The circumstances made for strange bedfellows. In his pocket book Llewelyn recorded how he persuaded a local radical to join an armed posse out hunting for rioters:

> I told him that I agreed with him that there had been grievances in Carmarthenshire which ought to be remedied & that I would be foremost to assist in any reasonable reform but that the law must be vindicated & the present lawless put down promptly before they had grown to dangerous proportions.[11]

His brother-in-law, Matthew Moggridge, also a magistrate and one of the local secretaries at the Swansea meeting as well as another leading light in the Royal Institution of South Wales, led similar sorties against the rioters.

The BAAS leadership must have been well aware of these recent events. Maybe they were one of the things they wanted Phillips to investigate during his visit. To have the British Association visit a part of the country where armed rebels were on the loose would have been unfortunate to say the least. But for civic leaders like the Groves and the Dillwyns getting the BAAS to town, and suppressing riots, were different sides of the same coin. Throughout Wales, new civic elites were on the rise and they wanted to transform the country in their own image. They wanted Wales to be part of the empire, not apart from it. That meant maintaining law and order, and it meant advertising their cultural credentials by persuading the BAAS to visit as well. It also offered a way of bringing the town together again. Radicals and reformers might be on different sides of the barricades as far as politics were concerned, but they could still rally around the banner of science. Getting the BAAS to town was a way of boosting Swansea's credentials as the 'metropolis of Wales' as well as offering a vision of how Swansea and Wales should take part in the imperial march of progress.

In practical terms, there was a great deal to do in the run-up to the BAAS's arrival – not least the business of collecting money. Newspaper advertisements soliciting funds for the occasion were addressed to the 'Nobility, Clergy, Magistrates, and all other Inhabitants of the Principality, who intend to contribute to the FUND for defraying the Expenses of RECEIVING the BRITISH ASSOCIATION'.[12] The published lists of subscribers suggest that even if most of those who contributed to the funds were from Swansea, there was at least some response from further afield, with subscriptions coming in from Cardiff, Carmarthen and Haverfordwest, for example. The BAAS's own leadership were clearly keen to promote the visit as a national event. Addressing the local committee a month before the meeting, John Phillips suggested that 'the expectation that the approaching meeting might be regarded as truly a meeting for Wales' was entirely justified, as he offered advice 'with the sole object of rendering this, the first visit to Wales of the Association, creditable to the town of Swansea, to the neighbourhood, and to the Principality'. His description of the forthcoming visit as 'the most important event in the recent history of the Principality' was fully in keeping with his hosts' ambitions.[13]

As well as collecting money, the organisers needed to make sure that there were enough places for visitors to the town to stay. Distinguished guests had no cause for concern in that respect, of course. They would be wined, dined and provided with comfortable beds by the civic leaders themselves. At various nights during the meeting, for example, Lewis Weston Dillwyn seems to have had the bishop of St David's, the geologists Henry de la Beche, Leonard Horner and Gideon Mantell, David Brewster, Edward Sabine, and even Prince Metternich, staying or dining.[14] More humble visitors were put up in hotels, or with the local householders who had promised accommodation for up to 200. Getting there was a problem to be overcome as well. Isambard Kingdom Brunel's Great Western railway had not yet arrived in south Wales in 1848; the South Wales Railway would not be completed until a few years later.[15] Travellers coming to Swansea for the BAAS meeting had the choice of either taking the mail coach along the south Wales coast, or of travelling by coach or train to Bristol and going by boat from there along the coast to Swansea. Bringing all of these arrangements together was no mean feat, and Grove and his fellow organisers must have been kept quite busy. They must have been quite relieved when 9 August dawned, with all their preparations in place.

A Scientific Meeting

The BAAS's first visit to Wales did not start well, in terms of the weather at any rate. It rained. 'The morning of the day upon which the British Association will hold its first meeting in Wales opens gloomily', complained the *Cardiff and Merthyr Guardian*'s correspondent,

> The rain pours down in torrents and has a depressing influence upon the prospects of this interesting occasion. – one to which the

> country has looked forward with that interest which ever attaches itself to events which, from their importance, become epochs in the life of a nation, in the history of a people or a race.[16]

Another newspaper noted how, on the meeting's first day,

> a large number of persons, including many distinguished visitors, as well as the gentry and principal inhabitants of Swansea and the neighbouring districts, availed themselves of the opportunity of becoming members of the Association ... When the reception room was closed on Wednesday evening, the total number of members enrolled was 575, of which 149 were ladies.[17]

Accommodation for the meeting was scattered throughout the town. Headquarters, predictably enough, were at the Royal Institution of South Wales's impressive building, which also played host to the chemistry section. The mathematics and physics section was in the girl's school across the road, whilst geology was housed in the town hall a few streets away. Someone with a sense of humour had placed the statistics section in the card room at the assembly rooms.[18] The organisers laid on squads of uniformed runners to carry messages between the venues across the town.

The meeting's first official event took place on the evening of the first day, to receive the council's report for the previous year and to confirm the appointment of officers for the meeting. Grove was amongst them, appointed vice-president along with Lewis Weston Dillwyn and John Henry Vivian from amongst the Swansea men. Then came the moment that Grove must have been dreading – the president's address. Either by chance or malicious design, the president for that year was no less a figure than the marquis of Northampton himself, president of the Royal Society and the main victim of Grove's own recent coup. What was he going to say? Northampton made it quite clear to his audience what a favour it was for the BAAS to visit a location so 'remote from the metropolis, remote from the chief seats of English learning, remote also from those great highways of communication by which modern

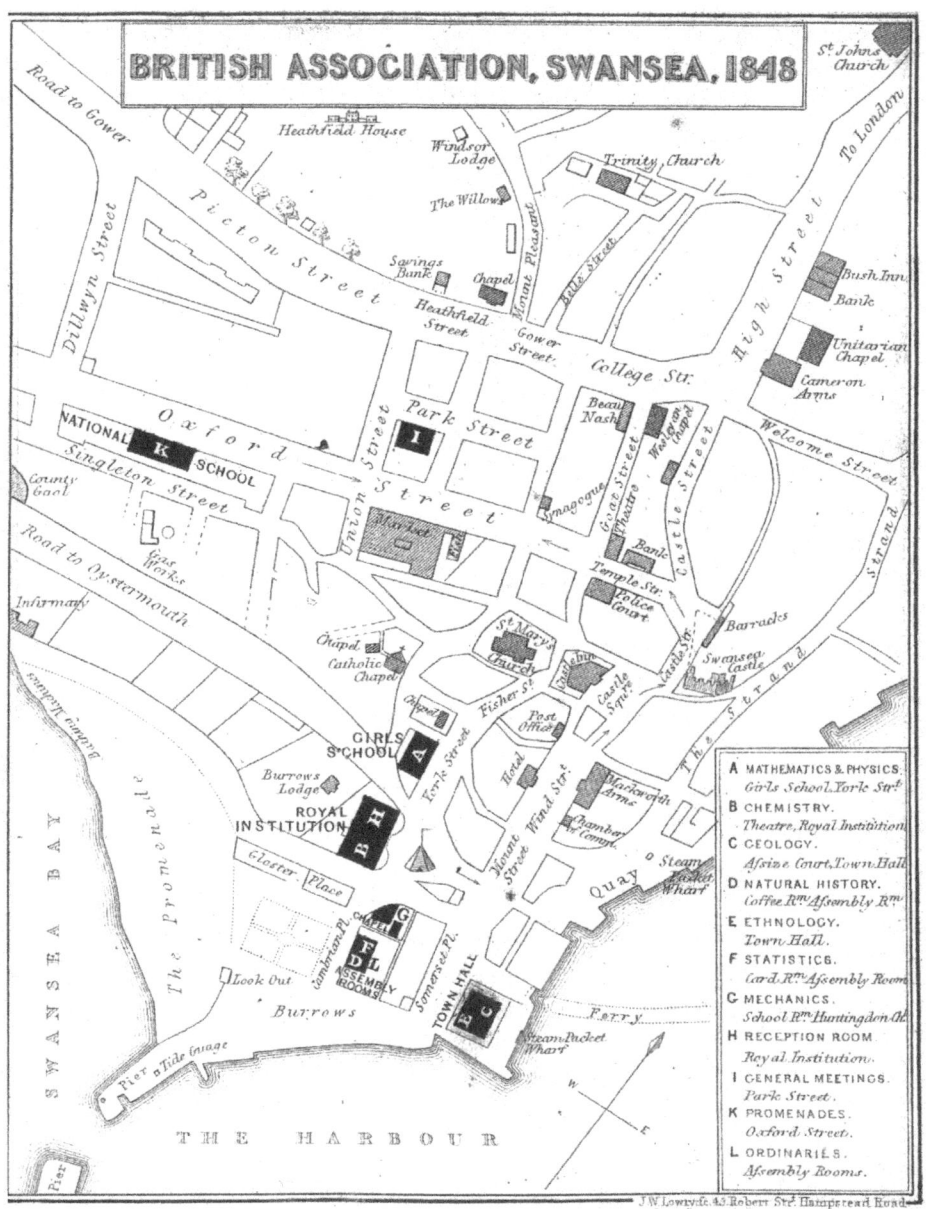

FIGURE 4 Map of locations for the Swansea meeting of the British Association for the Advancement of Science, 1848.

ingenuity has almost accomplished the extravagant wish of annihilating space and time'.[19] It was clear from Northampton's speech with its comparison with the previous year's meeting at Oxford, that in his view, at any rate, the BAAS was in Swansea to instruct, rather than be instructed. Oxford was a place where 'during the lapse of many centuries, science and learning have made their abode, and where religion has consecrated their union'.[20] Swansea, by implication, was not. It had no native talent to compare. It was 'in a corner, as it where, of Great Britain' and 'separated from the highways of steam'.

Northampton frankly warned his audience not to expect too much of the meeting: 'Swansea cannot with reason expect a meeting as numerous as those of York, and Cambridge, and Oxford, and still less like those that have congregated at Liverpool and Glasgow'.[21] So why were they there?

> To those members who were at Southampton and Oxford it would be quite superfluous to allude to the eloquent terms in which the advocate of Swansea, Prof. Grove, like a potent magician, or like a representative of the Bard and Druid of Ancient Britain, summoned us to the shores of the Bristol Channel.[22]

The jibe was clearly meant to remind his listeners of Hotspur's reply to Glendower's boastful 'I can call spirits from the vasty deep', that 'so can I, or so can any man; But will they come when you do call for them?' The BAAS had come, and only Grove's blandishments could have persuaded sober Englishmen to come to such a place, presumably just as only his glib tongue had persuaded the fellows of the Royal Society to give Northampton the boot. He dourly remarked that his only claim to the BAAS presidency for the meeting was his presidency of the Royal Society, and he was about to relinquish that. Grove himself, when he rose to thank Northampton after the address, presumably did so through gritted teeth.

But with the speechifying out of the way, there was plenty to keep everyone occupied through the coming days. The physical and mathematical sciences section, ensconced in the girls' school on York Street, featured a number of papers by the veteran Scottish natural philosopher David Brewster, on his favourite topic of vision and illusion. The audience at his discussion of Bishop Berkeley's theory of vision had the added entertainment of witnessing a head-to-head debate with William Whewell at the end of the paper.[23] In the same tradition of fascination with optical illusions, local man Matthew Moggridge, Dillwyn's son-in-law and a member of the local committee, gave an account of two cases of unusual atmospheric refraction. Moggridge described how at 'midday the 27th of January last we saw a schooner which appeared erect and resting on top of the high sand-hill east of the Neath River, the whole of the hull being visible'.[24] It was a mirage, of course. Amongst others, the Mancunian experimenter James Prescott Joule, was at the meeting too. He offered the section a further defence of his ideas about the mechanical equivalent of heat – a topic in which Grove would certainly have been extremely interested – and on the basis of 'a new and extensive series of experiments' offered a more accurate estimate of the equivalent.[25]

Grove presumably spent most of his time at the chemistry section, housed in the main theatre of the Royal Institution of South Wales itself, since he was section president. He gave his own performance there too, with a discussion of the 'peculiar cooling effects of hydrogen and its compounds in cases of voltaic ignition'. This was very much in keeping with his ongoing concerns with investigating the technologies of display. He performed an experiment for his audience that showed how 'a platina wire, rendered incandescent by a voltaic current, was cooled far below the point of incandescence when immersed in an atmosphere of hydrogen gas'.[26] Those attending the chemistry section could also have heard two papers by the Scottish engineer James Nasmyth: one on a 'peculiar property of coke' and another on 'the chemical character of steel'. Nasmyth described how coke was 'possessed of one of the most remarkable properties of the diamond, in so far as it has the property of *cutting glass*'.[27] Grove, presumably, would have been particularly interested in two papers by W. S. Ward: one describing a new design of galvanometer, 'in which a coil of wire conducting the electric current, was suspended around the poles of a U-shaped permanent magnet', and the other describing a series of experiments with the instrument.[28]

Over at the geology section, housed in the assize court in the town hall, there was plenty of opportunity for local men to make their presence felt. Spence Bate offered an account of 'fossil remains recently discovered in Bacon Hole, Gower; also other remains from beneath the bed of the river Tawey'. His finds included the

> teeth of the ox, deer, and other ruminant, together with a portion of the cranium of a deer and a few bones of a bat ... Teeth of carnivorous animals were also found, among which were the under left canine of an old *Ursus Spelaeus*; also the canine and molar of a young bear of the same species.[29]

Starling Benson, who was one of the section vice-presidents too, offered a couple of papers on the local coalfields. He described the geology of the coal deposits across south Wales, between Pontypool and Kidwelly, mapping the seams of coal of different quality. The different qualities were not strictly demarcated, he observed, but

> on the contrary, there is often a gradual change from bitumous to free-burning within the limits of the same colliery, whilst the free burning coals would also appear to become culms, burning without flame, probably from the diminution of volatile matter, before the quality of the true anthracitic coal and culm is attained.[30]

Those hoping to see one of the geological lions in action would have flocked to listen to Henry de la Beche discussing the geology of south Wales, Gloucestershire and Somersetshire.[31]

Lewis Weston Dillwyn presided over the natural history section over at the assembly rooms. He clearly took his presidential duties seriously, noting in his diary that he 'Attended all day as President of the Nat His Sect.'[32] The local lawyer John Gwyn Jeffreys – himself another Swansea FRS – gave a paper there on molluscs. Matthew Moggridge also gave a presentation here as well as to the physical science section, this time on a peculiarity of the algae protococcus nivalis that he had observed growing both on the Gower peninsula and on the banks of the river Pyrddin, on the edges of the Brecon Beacons.[33] Thomas Williams, a local doctor who had trained originally at Guy's Hospital in London and lectured on anatomy there, gave a lecture on 'the physical conditions regulating the vertical distribution of animals in the atmosphere and the sea'.[34] Williams had also lectured at Richard Grainger's Webb Street Anatomy School and had assisted Grainger when he performed a public dissection on the body of the political radical Richard Carlile in 1843.[35] Again, anybody attending in the hopes of seeing one of the big names performing would not have been disappointed. They could have listened to Edwin Lankester give an account of 'some vegetable monstrosities illustrating the laws of morphology'.[36] Or they could have listened to Richard Owen, the Hunterian Professor at the Royal College of Surgeons, talking about comparative anatomy and discussing the peculiarities of crocodile skulls or the importance of homology as a method in comparative anatomy.[37]

There was plenty on offer in the other sections too. Professor Elton offered the ethnology section, meeting in the town hall, a lecture on the 'ante-Columbian discovery of America'. After discussing the Vikings, he turned closer to home and described how three Welsh bards 'wrote before the discovery by Columbus, that Madoc sailed from Wales in 1170, and after pursuing a westward course for some weeks, arrived on a continent where the inhabitants differed from Europeans'.[38] Archdeacon Williams offered a discussion on 'the Gael, Breton, and Cymry'. Over at the statistics section in the assembly rooms' card room, presided over by Vivian, Cadogan Williams discussed 'the desirableness of extending to the working classes the opportunity of purchasing deferred annuities, as a provision for old age'.[39] Joseph Fletcher, secretary of the Statistical Society of London, offered some 'statistics of Brittany and the Bretons'.[40] Over at the mechanics section in the School Room on Huntingdon Road, behind the Royal Institution, the audience had the opportunity of viewing 'an artificial leg, of an improved construction', exhibited by a Mr J. Ashman.[41] More interesting to Grove would have been Francis Whishaw's exhibition of a 'subaqueous rope for telegraphic and other purposes'. This consisted of 'several small tubes of gutta percha, enclosed within a larger tube of the same material'.[42]

Entertaining Science

The scientific sections were only one part of the BAAS's activities during their annual meetings nevertheless. Just as important to almost everyone were the scientific entertainments. Evenings throughout the association's visit to town were dominated by a series of events, formal and informal, for the visitors' amusement – and to offer the locals an opportunity for rubbing shoulders with the distinguished guests. The *Welshman* thought that 'the arrangements connected with this department were judicious, and the dinners were served up in excellent style' At Thursday's dinner,

> the chair was occupied, as on the previous evening, by the Marquis of Northampton, who was supported on his right by Sir R. H. Inglis, Bart., J. H. Vivian, Esq., M.P., Sir J. J. Guest, M.P., T. W. Booker, Esq., High Sheriff for the county of Glamorgan, Sir H. De la Beche, &c. &c. On the left we observed the Bishop of St. David's, Lord Adare, M.P., Lieut-Colonel Sykes, &c. &c. There were also present, with many other distinguished individuals, the Dean of Westminster, Sir T. D. Acland, Admiral Sir Charles Malcolm, Dr. Rogers, Professor Grove, Leonard Horner, Robert Hutson, Esquires, H. Gwyn, Esq., M.P. besides most of the leading gentry and many of the clergy of the town, and the neighbourhood districts of the county.[43]

The big day for entertainments was Saturday, when the BAAS hordes headed out of town on a variety of organised excursions. One cavalcade headed up the Swansea valley to a variety of destinations industrial and bucolic. One division headed for James Palmer Budd's ironworks at Ystalyfera. After visiting the works they proceeded up the valley to visit

William Robert Grove

Figure 5 A Soirée at the Swansea meeting of the British Association for the Advancement of Science, drawing by John Weir Padley. National Library of Wales.

the Anthracite Mines of Abercrave; the Caves and Waterfall *Dan yr Ogof*, the Limestone hill of Cribbath; the Waterfall of Ischoed-yr-Henrhyd; and the sites of Sigillaria stems, one of which is now in situ in the rocks of the valley of Cwm-llech, and two of which have been removed thence, and are now preserved in the grounds of the Royal Institution, Swansea.

According to the *Welshman*, after dining 'under a long tent, erected in a field close to the river', the party divided with some going to 'see the anthracite mines of Abercrave, some to the caves and waterfalls in the neighbourhood'. The mines attracted the biggest crowd, but after going underground 'very few who entered will be induced to make another amateur visit to a coal mine'. Whilst all this was going on, the marquis of Northampton and Josiah Guest led their own exhibition to visit the excavations at Neath Abbey.

Another crowd went off the explore Carreg Cennen castle, built by the princes of Deheubarth in the twelfth century on a spectacular crag overlooking the valley, where they 'enjoyed the grandly picturesque scenery'. It was a trip of contrasts. As the *Welshman* noted in its report, all up the valley of the Swansea river the views are remarkably beautiful, the valley through which the river runs being a rich cultivated plain, and bold mountainous hills rise to a height of 800 feet direct from the valley. In the immediate neighbourhood of the town, however, the copper works emit masses of smoke and vapour that are not to be equalled in quantity by any manufacturing spot in England but, fortunately the smoke of the anthracite looks like white vapour, otherwise that part of the valley would be in darkness.

When the weary travellers returned to town,

> there was a promenade and soiree in the National School Rooms, which were fitted up for the occasion, and numerous objects of interest were hung round the walls and placed on the tables The bodily comforts of the members were well attended to by the local committee, who had provided refreshments on the most liberal scale.[44]

Another day trip set out to explore the 'Caves and Cliffs of Gower'. Those who did not wish to travel too far had choices closer to home as well. John Dillwyn Llewellyn opened the extensive grounds of his estate at Penllergaer to the BAAS visitors 'between 12 and 5 on Saturday'. As a particular treat they could see a 'Boat, impelled by the Electrical Current ... at work on one of the Lakes'.[45] Grove would have had a particular interest in this display since it was, of course, his own nitric acid batteries that powered the boat. It seems clear, too, that he had a hand in the design of the boat and its electromagnetic motor too.[46] The motor had been built by Grove's friend Benjamin Hill, and was probably similar, if not identical, to the one Hill described in a communication to the London Electrical Society in August 1841. In that communication he noted that

> I have constructed a working model of the machine here described, and which, with four small pairs of Grove's batteries, revolves with

sufficient power to turn small articles, although, from the necessary imperfection of the workmanship (being made by myself, with common carpenters' tools), the friction is excessive.[47]

Lewis Weston Dillwyn's diary entry for 19 August noted the departure of the last of their guests '& thus our Gaieties ended'.[48] It seems clear that the meeting's organisers felt that the BAAS's visit had been a great success. Grove had come out of it well too. He had been feted by his fellow townsmen for his part in the triumph. As the *Merthyr Guardian* noted:

> We must not pass unnoticed a very high (but withal most deserved) compliment which the inhabitants of Swansea paid to W. R. Grove, Esq. F.R.S., Barrister-at-law. This gentleman was entertained at a public dinner on the evening of Thursday week, – the chair being taken by L. Llewellyn Dillwyn, Esq., Mayor of Swansea, and the vice-chair by Starling Benson, Esq. A most influential and respectable party assembled, among whom but one feeling seemed to prevail, that the immediate object of their respect and esteem was eminently deserving of any compliment which it was in the power of his fellow-townsmen to give.[49]

He had seen his nitric acid battery in action on the lake at Penllergaer too, and his prominence in the organisation of the Swansea event can have done his prestige on the metropolitan stage no harm either. The meeting seemed to have consolidated Grove's position – and Swansea's place – in a network of mutual scientific patronage and exchange that bridged metropolis and province.

6

UNIFYING SCIENCE

With his Swansea triumph fresh in his mind, Grove was clearly anxious to continue the fight on his return to London. Reforming the Royal Society's statutes to limit the number of fellows that could be elected annually was the beginning, not the end, of his ambitions. Even before their reform proposals had been passed by the society's council in the teeth of fierce opposition from the marquis of Northampton and Peter Mark Roget, Grove and his fellow reformers were already plotting their next move. They were confident that their campaign had enough momentum to keep on going and that they should take full advantage of the opportunities offered by their enemies' disarray before the old guard had a chance to rearm and regroup. At Grove's initiative, the reformers had started to sound out a number of (hopefully) like-minded fellows with a view to establishing a society within a society. The possibility of setting up some kind of organised forum dedicated to the cause of scientific reform had been floated as early as February 1847. Towards the end of that month Leonard Horner wrote to Grove telling him that 'Owen, Murchison and Lyell heartily approve of the plan of your Club, and will be happy to join'. Grove and his allies were clearly already canvassing for support.[1]

If Grove's idea was to work, he and his fellow reformers needed to recruit as many potential supporters as possible into their camp. Having Richard Owen, Roderick Impey Murchison and Charles Lyell – leading and influential metropolitan gentlemen of science – on board was certainly a good start. Leonard Horner was despatched to try and recruit that old campaigner Charles Babbage, but 'made no impression on him'.² The lack of support from one of the leaders of the last reform battle two decades previously speaks volumes as to the ways in which what scientific reform meant had been transformed since the 1820s. John Peter Gassiot, Grove's friend, patron and fellow electrician, was given the task of trying to persuade that other veteran of the last campaign, John Herschel, to join them. He explained to him how, for

> some time past several of its sincere wellwishers have considered that the Royal Society might derive much advantage provided a Club could be constituted, <u>strictly limited to Fellows who have in some distinct manner contributed to scientific research</u>. Such a Club to be moderate in scale of expenditure and meeting twice a month during the sittings of the Royal Society, to ensure a fuller attendance of scientific Members at our Ordinary Meetings.³

Herschel, another old stalwart of the 1820s reform campaign and the reformist candidate against the duke of Sussex in 1830, turned out to be as difficult to persuade as Babbage. He was worried that the proposed new body would trespass on territory occupied by the already existing Royal Society Club, and Gassiot had to reassure him that many of the initiative's supporters were themselves leading members of the older body. Herschel eventually allowed himself to be persuaded and allowed his name to be listed as one of the club's founding members. Nevertheless, he was not one of the twenty-seven fellows at the club's inaugural meeting on 12 April.⁴ It was held at Clunn's Hotel in Covent Gardens – a popular location for literary dining and drinking clubs. The Fielding Club founded by William Makepeace Thackaray met there a few years later, for example. One of the first issues to be discussed at this meeting was the thorny question of a name. One proposal was that the new body should be called the Forty Seven Club, to mark both the year of its establishment and the triumph of the reformers. Eventually, however, the meeting agreed that it would be called the Philosophical Club. The name offers an interesting insight into the way in which Grove and its founders saw themselves in relation to the Royal Society and the other metropolitan scientific societies.

The Philosophical Club's ostensible purpose was to

> promote as much as possible the scientific objects of the Royal Society, to facilitate intercourse between those Fellows who are actively engaged in cultivating the various branches of Natural Science and who have contributed to its progress, to increase attendance at evening meetings, and to encourage the contribution and discussion of papers.

The club's membership would be limited to forty-seven at any one time, as a reminder of the year in which it had been founded and the Royal Society's statutes reformed. With the exception of the Royal Society's president (and it speaks volumes of Grove's and the other founders' ambitions that they regarded future presidents as obvious Philosophical Club fodder), only fellows who were 'authors of a paper published in the Transactions of one of the Chartered Societies, established for the promoting of Natural Science, or of some work of original research in Natural Science', would be eligible to join. The meeting elected Grove as treasurer (the only named officer) and the geologists Leonard Horner, Edward Forbes, the chemist Thomas Graham and the botanist John Forbes Royle as his Committee of Management. The committee was authorised to recruit additional fellows into the club until the full complement of forty-seven had been achieved.[5]

The Philosophical Club was soon packed with influential men. Of the forty-seven founding members, twelve were serving members of the Royal Society's council, including three of the society's officers. They also included many officers and council members of other metropolitan scientific societies as well as sixteen members of the officers and council of the British Association for the Advancement of Science (BAAS) for 1847. Leading members of the Geological Society in particular would play prominent roles in the club's activities during these early years. Charles Lyell joined his father-in-law Horner as a vocal supporter, and Henry de la Beche with his Swansea connections as well as Edward Forbes and Roderick Impey Murchison were active too. Horner predicted optimistically that future historians would be able to see that 'great and important changes in the Royal Society and the progress of science can be traced to the institution of the Philosophical Club'. But he also recognised that if that prophecy were to be realised, 'there must be much determination in those who take the lead in the Club to lose no time in laying the foundation on which we are to build our hopes of a new state of things'.[6] Edward Forbes hoped that the metropolis's other scientific bodies would learn to 'look at the Philosophical as a sort of higher council or guardian angel of them all'.[7]

The members of the 'higher council' were to meet monthly between October and June, on a Thursday, so that they could then proceed after dinner to the Royal Society's monthly meeting. At meetings of the club, members were invited to 'bring forward any correspondence or scientific subject worthy of consideration' by the assembled gathering. Topics raised for discussion during the first few meetings and recorded in the club's minute book ranged from the desirability or otherwise of adopting the centigrade scale for measuring temperature to the sightings of giant sea serpents off the coast of North America. Interspersed with those sorts of discussions were conversations about the current state and future prospects of the Royal Society. That, after all, was what the club was meant to direct.[8] Grove and his fellow reformers wanted the Philosophical Club to act as a forum for working out the future of British science. More immediately, in the aftermath of the successful implementation of the charter committee's reforms to limit the fellowship, and the announcements by both Northampton and Roget that they would resign their positions, they saw an opportunity to further tighten their grip on the Royal Society. They were determined that it would be the Philosophical Club who would decide who the next president and the next secretary of the Royal Society would be.

Election

Who should be the philosophical candidate to be president of the Royal Society? The geologist Edward Forbes made the job description quite clear in a letter to Grove towards the end of March 1848:

We must if possible have a scientific president – not a peer – or there will be a revolt and barricades constructed scientifically in the antechamber of Somerset House. Now or never is the time and if the Council won't do what it ought I hope the members in mass will rebel and demand a new constitution. Science demands a republic. Hurra![9]

The revolutionary metaphor was quite apposite, of course, in that year of revolutions, but Forbes was not just being facetious. Grove and his friends really did think that what they planned to do to the Royal Society would be revolutionary and would have wider repercussions for the Victorian world beyond the society's meeting chamber in Somerset House. Michael Faraday was suggested as a presidential candidate that would fit the philosophical bill, but he declined on grounds of ill health. They considered John Herschel as well, but worried that he lacked stamina for the trials ahead. 'Herschel will be far better fitted to take the chair in settled times, than he would be while changes & reform are in progress', Horner told Grove.[10]

In the end the Philosophicals united around the earl of Rosse as their official candidate for the presidency. He had the advantage of being a peer (despite Forbes's prejudices) and therefore being more likely to be acceptable to the old guard, as well as being scientifically active. Rosse was a well-regarded astronomer, famous for having had built the Leviathan of Parsonstown – a gigantic telescope – at his family seat in Ireland.[11] In a further coup, they also succeeded in persuading the council that the presidential term should be limited. At the 13 April council meeting Henry de la Beche put forward a resolution 'that in the opinion of this Council, it is inexpedient that the name of the same Fellow of the Royal Society should be inserted, as President, in the House-list of the Society for more than four successive years'.[12] As far as the presidency went, the Philosophicals had therefore seen their man chosen as the council nomination, and set a precedent for limiting the presidential term of office. The matter of the secretaryship turned out to be considerably more controversial. Grove himself would be the philosophical candidate for the position, but he and his allies were painfully aware that it would be a very difficult campaign.

Henry de la Beche wrote Grove a long letter towards the end of April, expressing his misgivings. 'After parting with you last night,' he wrote,

> I began seriously to reflect how far friends of yours were justified in placing one willing to do so much for the progress of the R.S. in a position that might turn out to be a false one, caused all kinds of bother and strife – with heaven knows what result. Here you are a man distinguished alike for your scientific advance and progress at the law, and having every right to look forward to professional advancement, brought into contact with things and people not very pleasant – and after all with all the chance of sacrifice on your part. There can be little doubt of your importance as Sec^y. for the advancement of that progress so much needed at the R.S. but the sacrifice would be less if you were Foreign Sec^y. – as I gleaned from you last night you would prefer.[13]

They knew that there was every possibility that Grove's candidature might not succeed and that his many enemies might see it as an opportunity to strike back and put this upstart Welshman in his place. Grove himself was clearly in two minds about the prospect.

Grove's main difficulty was that his attempt to gain Roget's soon-to-be-vacated position flew in the face of a long-established Royal Society tradition that the two secretaryships should be divided between physiology, broadly speaking, and the physical sciences. Roget was considered a physiologist, and the other remaining secretary, Samuel Hunter Christie – a Trinity College, Cambridge-trained mathematician and electrical experimenter – was considered to represent the physical sciences. If Grove stood he risked alienating the Royal Society's substantial contingent of medical men – and he and his allies could be sure that Roget would take full advantage of the opportunity that offered to take revenge for his own downfall. Much backroom politicking took place before Grove was eventually nominated as the council's official candidate for the secretary's position at their meeting on 6 July. Lyell complained bitterly about the difficulties involved in bringing matters to a successful conclusion 'against a set of obstructives, compared to whom Metternich was, I presume, a progressive animal'.[14] When it was all done he wrote to Grove to congratulate him on his nomination, commenting that 'We have narrowly escaped a shipwreck of all you and Horner and others had been doing for 3 years.'[15] But would it really be plain sailing from then on?

Grove must have felt some confidence as he left for Swansea and the BAAS. The society's procedures meant that as the council's official candidate his name would be the only one on the ballot. Anyone wishing to vote for another candidate would have to scribble out his name and write another in instead. He must also have known, though, that trouble was brewing. There had been rumblings already of meetings to organise another candidate. When he and his fellow Philosophicals returned to town following Grove's triumph at Swansea, it was to discover that the rumblings were quite true. Roget had gathered a committee to lobby for the election of Thomas Bell to the vacant secretaryship. An old hand at the dirtier side of Royal Society politicking, Roget knew exactly how to approach this kind of fight. The choice of Bell was a clever one. He was a fellow member of the Philosophical Club (which must have hurt) and his candidacy allowed Roget to disguise the battle as one between the physical and the natural sciences rather than as a set to between reformers and opponents of reform. The fight to get Grove's name onto the official list of council candidates would now have to be fought all over again – and this time in a far more public forum.

Roget's allies soon unleashed a fierce press campaign against Grove and the reform camp. An anonymous correspondent styling himself 'A JUNIOR F.R.S.' wrote to the *Athenaeum* asking why there were so few practitioners of the natural sciences (i.e. the life sciences) on the council list: 'is it that there are fewer naturalists than chemists and mathematicians who have been able to pass through the second ordeal of the ballot box of the Philosophical Club, more terrible than the first in the Society?'[16] Another pointed the finger straight at Grove. He had been 'one of the originators of the retrograde reform contrived to prevent the diffusion of knowledge and, if possible, return us to the exploded exclusiveness of the Dark Ages'. If that was not enough, 'his conduct is somewhat too imperious to hold so important a situation; where he has to deal with his equals at all times, – with his superiors at most'. There were 'several Groveites, or retrograde reformers' on the council list too, this correspondent averred. Some of them were 'but schoolmasters with another name, – and the greater portion are selected exclusively from a small department which decreed, with the imperiousness of an Oriental Nabob, that Physics are all things, Phytology nothing'.[17]

Not all the press was negative. The *Literary Gazette* editorialised that if Philosophical Club cliquery was the issue, then 'to such a working clique we would willingly entrust the interests of Science and the Society, until "Fellowship" be restored to a healthy state, leaving jealousy and discontent to those who have too long traded with F.R.S.'[18] The *Gazette* averred that 'we espouse no party in this matter, but we earnestly advise scientific men each to sacrifice a little of his own opinion, or what he may conceive the interests of his own branch, to the general good of science'.[19] The radical medical journal, the *Lancet*, was less circumspect.

> We look on these proceedings of the Physiology Committee, not as a vindication of physiology, but as an attempt to annul the radical changes which ought to follow on the retirement of Dr. Roget from the secretaryship, and which should involve the destruction of the Physiological Committee, and the other secret and irresponsible committees, all of which are pests of the Royal Society.

This was not a battle between physics and physiology, whatever Roget and his camp might pretend: 'the question at present at issue is not, to our apprehension, the ejection or neglect of physiology, but whether the Physiological Committee, which has disgraced physiology, shall be permitted to beard the reforming council?'[20]

Many potential allies were clearly torn. William Bowman, one of the rising stars of British anatomy, wrote to Grove telling him that it was 'with very great reluctance that I have come to the conclusion to give a vote on St. Andrew's day, which may have the appearance of being in opposition to your great and just claims to the office of Secretary to the R. S.'[21] The up-and-coming physiologist William Benjamin Carpenter worried that 'Two distinct questions are unfortunately mixed up in the present case. The first, whether the Society ought to have a <u>Physiological</u> Secretary. The second, whether it ought to have a <u>Reforming</u> Secretary.'[22] These were just the sort of ambitious and forward-looking men of science the reformers wanted onside, and they were clearly failing to recruit them. Roget's tactics of divide and rule were working. In the event, Grove's camp had completely failed to anticipate the power the marquis of Northampton would have to direct the proceedings – his last act before standing down as president. Presiding over the crucial anniversary meeting, Northampton announced that those who had copies of the unofficial lists naming Bell as candidate for the secretaryship could use them in the ballot – in direct contravention to the society's policy. They were to vote once for the council and then separately for the officers. This put Bell's supporters at an advantage since only they had copies of both the official and unofficial lists. It looked, as the *Literary Gazette* put it, 'part of a deeply laid scheme, or the suggestion of a subtle tactician'.[23] The result was a defeat for Grove by twenty-six votes.

Grove and his allies were predictably furious. An incandescent Leonard Horner wrote to him that it had been his

> full intention to have called upon you this morning at breakfast time to give vent to my feelings of disgust at the whole of yesterday's proceedings, so disgraceful to the Society, but most disgraceful of all to Lord Northampton, who has proved himself a man unworthy of all confidence. If we had for a moment imagined that there was in the RS a set of fellows so low as those who have gone through such dirty work as we have witnessed in the active men of Bell's committee, I am sure none of your friends would ever have asked you to allow them to put you in nomination.[24]

Charles Lyell told Grove that he 'would rather be beaten than let the organ of a party who are the enemies of mental progress walk over the course'.[25] They had to be satisfied that for the moment, at least, they had been successful in getting Lord Rosse as the new philosophical president. Grove himself had not taken kindly to defeat. He made no further efforts to gain office in the Royal Society. He remained active, as we shall see later, in the Philosophical Club, but never again would he agree to act as the public candidate for reform.

Investigating the Discharge

One obvious consequence of Grove's decision to resign his professorship at the London Institution back in 1846 was that he lost his laboratory. As a professor there he had access to space and resources that rivalled what was available to Faraday at the Royal Institution. He now had to look elsewhere for a place to experiment. There was certainly a tailing away of Grove's experimental activity by the end of the 1840s. Maybe the decision to take a more active role in Royal Society politics was, at least in part, motivated by a desire to find some other way to make a place for himself in science. If so, then by the end of 1848, despite his triumph at the Swansea meeting of the BAAS, Grove was in trouble. He was without a lab and without the influential role in setting the direction of metropolitan science that he had coveted. Despite all this, Grove did start a new episode of experimental work towards the end of the decade. Along with John Peter Gassiot he started investigating the mysterious glowing light that appeared inside sealed glass tubes when they were hooked up to an induction coil, or other similar source of high intensity electricity.[26]

In some respects, this was relatively familiar experimental territory for Grove. Much of his research over the previous decade, since the invention of his nitric acid battery in 1838, had been directed towards dissecting the familiar phenomena of the electricians' technology of display – sparks, electrical combustion and the like. We have already talked about some of this work in an earlier chapter. Grove had been concerned to investigate not just how best to produce the various phenomena of spectacular electricity but to make the whole process through which they were produced visible too. He did not just want to produce sparks – he wanted to get inside the spark too. Some of Grove's experimental work following his resignation seems to have been carried out in Gassiot's laboratory at his home in Clapham Common. The wealthy wine merchant could afford expensive and extensive apparatus. In his discussion in the *Philosophical Transactions* of 'The Effects of Surrounding Media on Voltaic Ignition', he mentioned that he had been able to repeat one of his experiments on a more extensive scale 'by the kindness of Mr. Gassiot' and making use of his 'battery of 500 well-insulated cells of the nitric acid combination'. It seems safe to assume that, given the size and fragility of the apparatus, Grove came to it rather than it coming to him.

Grove's 1849 publication offers a good place to start an overview of the experimental research that would intermittently occupy his attention over the next decade or so. As he observed, several years ago he had 'pointed out a striking difference between the heat generated in a platinum wire by a voltaic current, according as the wire is immersed in atmospheric air or in hydrogen gas'.[27] His 1849 paper described his attempts to study this phenomenon more closely. As Grove acknowledged this was, in part at least, a more speculative extension of the practical experiments he had conducted back in 1845 which led to his paper on the application of voltaic ignition to lighting mines. Now he was trying to understand what properties of the surrounding medium helped make the process of ignition work better. Why did the wire burn more brightly in some gases compared to others? Grove's 1852 paper on the electro-chemical polarity of gases continued this focus on the medium. He wanted to understand better the properties of the surrounding dielectric and work out whether there were circumstances in which it actually conducted electricity. In this he was frustrated by not having a battery of the 'enormous intensity' required.[28] Grove only succeeded in solving this problem when, during a visit to Paris, the French physicist César-Mansuéte Depretz drew his attention to a new piece of apparatus that might do the job.

This was a new kind of induction coil, constructed by the German instrument-maker Heinrich Ruhmkorff, then resident in Paris. Induction coils had been a standard item in the electrician's armoury since they were first developed by the Irish philosopher priest Nicholas Callan in the aftermath of Faraday's discovery of electromagnetic induction.[29] They consisted of two coils of wire – an inner one composed of a relatively small number of loops of thick copper, and an outer coil made of a larger number of loops of thin wire. When the inner coil was consecutively connected and disconnected to a battery a far more powerful current was generated in the outer coil. In effect it offered a way of generating an intense current from a small battery. By the 1840s they were particularly popular as medical electrical devices, being cheap to operate and quite easy to manipulate.[30] Ruhmkorff's achievement was to take this relatively simple instrument and turn it into a far more powerful piece of apparatus. This was done by improving insulation and winding the coils in such a way as to maximise the instrument's efficacy. Following his conversation with Depretz it did not take Grove long to get his hands on one of Ruhmkorff's instruments and to set about experimenting.

The apparatus Grove acquired was '6.5 inches long, 4 inches diameter; the length of the wires forming the coils are (I give M. Ruhmkorff's measurements) stout wire, 30 metres long, 2 millimetres diameter, 200 convolutions; fine wire, 2500 metres long, ¼ millimetre diameter, 10,000 convolutions'.[31] With it he could generate just the kinds of high intensity currents of electricity that he needed for his experiments on the electric discharge. He was interested in investigating whether gases could be conductors of electricity, and had been carrying out 'experiments on the voltaic arc taken in various gaseous media, with the view of ascertaining the state of the intervening media anterior to, during, and after the discharge'. With this in mind he was looking 'for some modified form of electric discharge which should be intermediate between the voltaic arc and the ordinary Franklinic discharge'.[32] He wanted to see whether the discharge produced any chemical effect on the gases through which it passed. He discovered that when the apparatus was arranged with one electrode being a plate and the other a needle, the discharge caused beautiful oxidised rings to appear on the plate. Grove speculated whether there was some relationship between these rings and the dark bands that he observed in the discharge itself.

Grove also lent his new Ruhmkorff coil to Gassiot, presumably returning the favour of having been able to use Gassiot's laboratory facilities when he needed them. Gassiot was soon carrying out his own experiments, culminating in the spectacle that would come to be known as Gassiot's cascade. In this experiment a glass cup lined with tinfoil was placed inside an air-pump with an electrode placed at its mouth. When the terminals of the induction coil were attached and the pump evacuated,

> at first a faint clear blue light appears to proceed from the lower part of the beaker to the plate; this gradually becomes brighter until by slow degrees it rises, increasing in brilliancy, until it arrives at that part which is opposite or in a line with the inner coating the whole being intensely illuminated. A discharge then commences from the inside of the beaker to the plate of the pump in minute but diffused streams of blue light; continuing the exhaustion, at last a discharge takes place in the form of an undivided continuous stream, overlapping the vessel as if the electric fluid were itself a material body running over … streams of lambent flame appear to pour down the sides of the plate, while a continuous discharge takes place from the inside coating.[33]

The striae, or dark bands, that Grove had observed in the discharge were a source of much speculation. Gassiot devoted his Bakerian lecture to them in 1858.[34] Grove returned to the topic himself a few months after Gassiot's lecture. He argued that the striae were the result of the intermittent current from the Ruhmkorff coil. To demonstrate this he devised an apparatus that introduced a spark gap into the secondary coil so that only single discharges could take place. With this modification,

> the coil apparatus was set at work, and the striae very beautifully exhibited in the receiver; the projecting wires were now gradually separated, and the stria for some time continued visible and until the points of the wires were so far apart that an occasional spark

only passed from point to point when the striae disappeared, and a uniform luminous cloud was produced in the receiver.[35]

Again, some of the apparatus he was using was borrowed from Gassiot. He noted that in some of the experiments he used

> one of Mr. Gassiot's vacuum-tubes, which showed the striae beautifully under ordinary circumstances; but when the division in the circuit was carefully made and carried to its fullest extent, the discharges passed without any striae, the tube being filled at each discharge with a uniform glow.[36]

It seems likely that by now his opportunities for experiment were so limited that it was no longer worth his while to invest in his own equipment.

Juxtaposition

Once he had regained his composure following the personally catastrophic election campaign for the Royal Society's secretaryship, Grove was soon campaigning again. Like many of his fellow members of the Philosophical Club, Grove was convinced that men of science needed firm leadership in order to guarantee continued scientific progress. As he put it to his fellow Philosophicals, 'the cultivators of Science suffer from the isolated character of the principal Scientific Societies that there was a want of "espirit de corps" and central authority among Scientific men and a consequent temptation to schism and encouragement of charletanery'.[37] The solution, according to Grove at least, was juxtaposition – an attempt to bring the fractured (and fractious) specialist societies back literally under the same roof as the Royal Society. This was clearly a matter that would need to be treated with some care. Grove emphasised that if juxtaposition 'were a desideratum it would be obvious that the more moderate the measures proposed to effect this, the more likely were they to be successful and the less likely to injure existing institutions'.[38] In the first instance, he suggested, members of

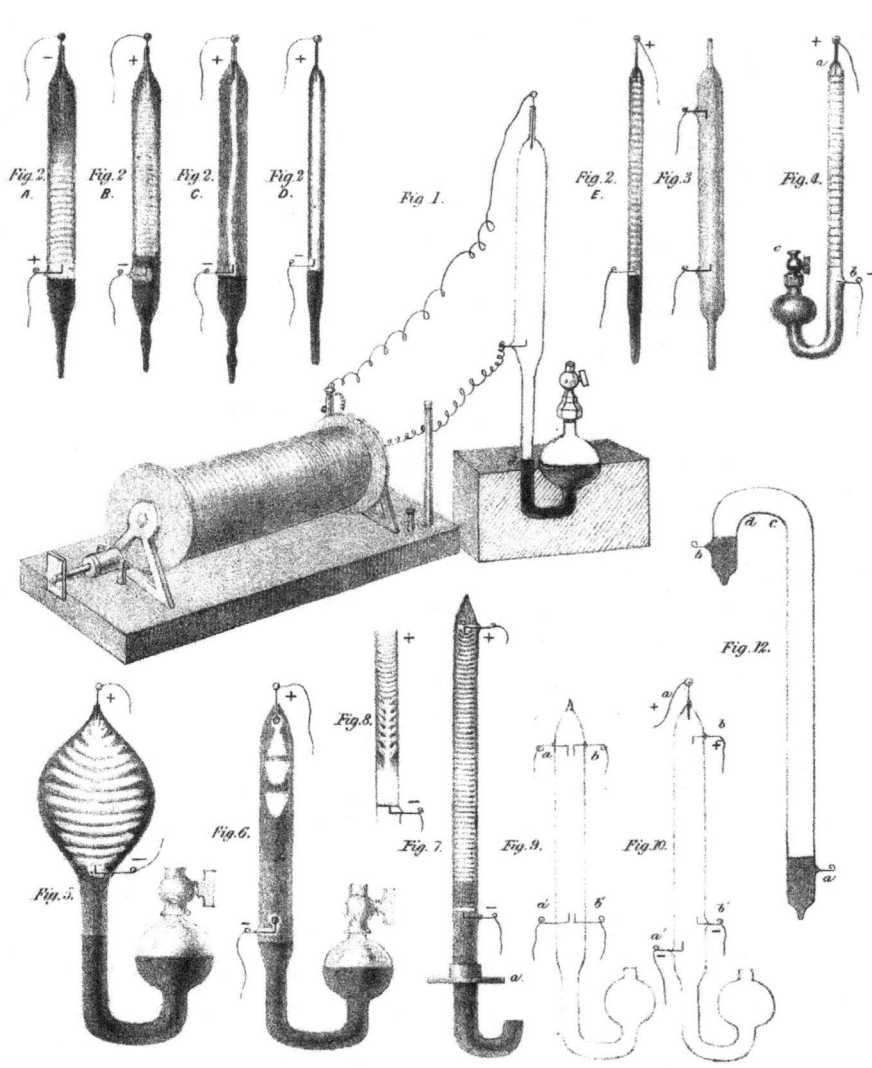

FIGURE 6 Discharge experiments, *Philosophical Transactions*, 1858.

the Philosophical Club who were also members of the councils of the various specialist scientific societies (like the astronomical, chemical, or geological societies) should 'exert themselves to obtain for such societies a single locale'.[39]

It would not be until 1852 that Grove and the Philosophical Club had a real opportunity to push for juxtaposition. There had been hints in 1850 that the government was considering offering some of the vacant London mansions they had at their disposal for housing some of the scientific societies. Grove, the anatomist Richard Owen and the Royal Society's foreign secretary Edward Sabine were instructed to investigate the possibility.[40] The chemist Lyon Playfair suggested that part of the building that would be built in Hyde Park for the Great Exhibition should be a permanent installation and made available for use by the scientific societies. Playfair's proposal found little favour. His fellow Philosophicals were not keen on the idea of moving to the outskirts of the city, away from the seats of social and political power.[41] During 1852, however, the Royal Society's own leadership entered into negotiations with the government in an effort to secure better accommodation for themselves. The Philosophicals immediately embarked on their own campaign of lobbying directed at Lord Rosse, the Royal Society's president, reminding him of the advantages offered by juxtaposition and urging him to raise the matter in his own negotiation with government. Efforts were made to solicit the support of the other scientific societies for the move.[42]

The following year, Grove and other leading Philosophicals prepared their own memorandum to be signed by representatives of the leading scientific societies and presented to the prime minister. At that year's anniversary meeting of the Royal Society (the scene of Grove's humiliation at the hands of his rivals five years previously), Grove was appointed part of a committee to deal with the matter. Lord Rosse, who was, after all, the Philosophicals' own candidate for the society's presidency, came out strongly in favour of juxtaposition in his presidential address. For 150 years, he said, the Royal Society had carried on its shoulders 'the whole labour of wielding the power of Association, in the cause of progressing science'. The emergence of the specialist societies during the early decades of the nineteenth century might have helped lighten the load, but had not changed the society's responsibility. Without the Royal Society's leadership, science would splinter, he argued. Rosse told his audience that specialism 'would proceed too far were there not a power to restrain it: you hold that power: you exercise a presiding influence over all the Societies. The leading members of the scientific bodies have their places here, and Science is fully represented.'[43] Grove might have written the script himself.

The Philosophical deputation met with the prime minister, Lord Aberdeen (himself a Peelite and leading a government made up of a mixed bag of Peelites and Whigs) on 23 May 1853. Aberdeen was sympathetic and promised to discuss the proposals with Gladstone, who was Chancellor of the Exchequer. A year or so later, the deputation met with the Home Secretary, Palmerston, for further negotiations. The issue of juxtaposition was raised again by the BAAS's parliamentary committee under the chairmanship of Lord Wrottesley – another member of the Philosophical Club and soon to be Rosse's replacement as president of the Royal Society. By now the Philosophicals had a specific building in mind – Burlington House – and were lobbying hard to gain possession. Eventually, in 1857 they succeeded in getting at least a limited version of their ambition. The Royal Society, the Chemical Society and the Linnean Society moved into new quarters at Burlington House that year. It was a limited victory, but significant for Grove nonetheless. The virtue of juxtaposition was that 'the advantages derivable from *concentration* would be combined with those derived from separate and independent action'.[44] It meant that Grove and his allies in the Philosophical Club could present themselves as a united front for science.

Grove was clearly still casting around for a role for himself in science. In 1855 Lord Wrottseley compiled a report on the BAAS's parliamentary committee's activities, and drew heavily on a letter from Grove. Grove had argued that 'scientific men have but very limited means of acting on Government, they are politicians to a lesser degree than any of her majesty's subjects, they consist of men belonging to various classes of society and whose ordinary occupations differ greatly'. In other words, science needed men like Grove to do the talking. They needed to forge a productive relationship with the Victorian state as well. Grove and Wrottseley argued that 'the great measures of reform or progress [which] are affected in this country result from strong pressure of public opinion urged on by agitation and as men of science are peculiarly unfitted for this process Government might not unreasonably be asked to step out of its usual habits and to lend science a helping hand'.[45] Around this time, Grove was also contemplating the possibility of putting his name forward for the chair of chemistry at Oxford that had just been vacated by Charles Daubeny. In the end he decided not to be a candidate and the chair was awarded to Benjamin Brodie. Grove had not wished to submit himself to a contested election – another legacy of his bruising at the Royal Society in 1848.[46]

7

A SCIENTIFIC STATESMAN

It is clear that by the end of the 1850s, if not sooner, Grove had abandoned any hopes of an active career in science. His half-hearted flirtation with the Oxford chair of chemistry was his last fling in that direction. Around about this time too, his evidence to the Parliamentary Committee of the British Association for the Advancement of Science (BAAS) also suggested that the 'establishment in the metropolis and perhaps in some of the large provincial towns of public professorships in different branches of science would be highly advantageous'. He presumably had himself in mind for one of these plum positions in the event of their becoming a reality. After all, he said,

> the pursuit would in itself be so tempting to the men most likely to excel in it, that no higher salaries would be requisite than such as would enable them to educate their children and to retain their position as honoured members of the middle classes of society.[1]

It was an increasingly forlorn hope on Grove's part, as he must have realised. In any case, by the middle of the 1850s his professional career in the law was flourishing as he started to make a name for himself as an up-and-coming barrister.

Grove had been called to the bar in 1835, but there is little indication of his activities as a barrister for the first fifteen years or so of his legal career. He must have had some practice since one of the reasons he gave for resigning his position as professor at the London Institution was that his increasing legal responsibilities meant that he no longer had the time needed to devote to his professorial duties. By 1853 he had been appointed a Queen's Counsel, itself a mark of at least some degree of eminence and success in his legal career. Having taken silk he soon established himself as a leading member of the Chester and the south Wales circuits. His legal career was obviously prospering, and despite his continued involvement in the Royal Society's affairs and occasional evidence of hankering after a scientific position, Grove must have realised that it was there that his hopes of advancement lay. What science he did from then on – and his dwindling list of publications makes it clear that it was increasingly little – would have to be fitted into any spare moments left over from his pursuit of the law. He must have been uncomfortably aware that he looked less and less like the sort of active man of science he had himself argued was the only proper candidate for fellowship of the Royal Society.

There were times when science and law coincided, however. That was the case with Grove's involvement with the case of the notorious William Palmer, the Rugeley poisoner, for example. Charles Dickens called Palmer, who had been charged with poisoning a friend with strychnine, 'the greatest villain that ever stood in the Old Bailey dock', and asked how anyone could ask for a display of sensibility from such a man.[2] It was Grove's task to defend him. Palmer, a medical practitioner with a passion for gambling on the horses had allegedly killed his fellow racing enthusiast John Cook with strychnine, in the hope of getting money to pay off his mounting gambling debts. Despite the inquest's failure to produce any physical evidence of poisoning, the jury found that he had died of poison administered by Palmer and a special act of parliament was passed to allow him to be tried at the Old Bailey. The defence was led by the Queen's Serjeant, William Shee and assisted by Grove. Grove had been chosen by the defence because they hoped his specialist chemical knowledge would help prove the case in Palmer's favour. Palmer was found guilty despite the evidence of a number of medical authorities who insisted that had strychnine been used then traces of it would have been found during the post-mortem, and executed at Stafford prison in front of a crowd of 30,000 witnesses.[3]

At about the same time, Grove was also involved in another case which, though considerably less gruesome, must have provoked some mixed feelings too. In 1855 the patent for the calotype process of photography that had been taken out in 1841 by William Henry Fox Talbot was due to expire. Talbot had to decide whether to allow the patent to lapse or take expensive legal action to extend it. At the same time, he was faced with a challenge to the patent from Martin Laroche, a professional photographer with a studio on Oxford Street and who had exhibited samples of his work at the Great Exhibition. Like many of his profession, Laroche was keen to break a monopoly that they saw as interfering with their business and he tried to do this by using a different chemical as the developing agent. Talbot, in turn, was determined to protect his patent and took Laroche to court. It was an important case for professional photographers who regarded Talbot's patent – as well as that of Daguerre, his French counterpart – not only as an unnecessary expense but also as a barrier to their own efforts to experiment and improve photographic processes. Even early efforts to establish a photographic society had foundered because of difficulties with Talbot's patent.[4]

Grove had links with Fox Talbot going back to his Swansea days. His contemporary John Dillwyn Llewelyn had married Emma Thomasina Talbot, Fox Talbot's cousin. The Talbot family were a major force in Swansea and Glamorganshire, both as established gentry and as important industrialists. That local connection, as well as his philosophical reputation and knowledge of photographic processes, may have been one of the reasons that he was called to assist Sir Frederick Thesiger, who acted as Talbot's leading counsel when the case was brought to trial. Grove had been involved with experimenting in photography since the early 1840s, at least. His work on using electricity as a means of reproducing daguerreotype plates, for example, was an attempt at combining his electrical and photographic experimental interests, and he noted how Daguerre's patent had made it difficult for him to obtain plates for his experiments.[5] Grove also discussed photography in his essay *On the Correlation of Physical Forces* and used daguerreotype plates in some of his experiments to illustrate correlation.[6] Talbot was similarly interested in Grove's electrical work. In the immediate aftermath of the Laroche case, Talbot discussed with Grove his work on the electrical discharge, telling him that he would like to see his 'experiment of the Ruhmkorff coil applied to the Leyden jar and with your assistance to apply it to obtain some instantaneous picture as I did formerly'.[7] Legal and experimental work could still coincide.

Grove was deeply opposed to the taking out of patents by gentlemen of science. As he expressed it in his attack on the state of science published over a decade earlier: 'It would scarcely add to the dignity of philosophy, or to the reverence due to its votaries, to see them running with their various inventions to the patent office, and afterwards spending their time in the court of law, defending their several claims.' As far as he was concerned, if 'parties look to money as their reward, they have no right to look for fame; to those who sell the produce of their brains, the public owes no debt'.[8] Grove had Fox Talbot directly in his sights when he wrote this. Commenting on his award of the Royal Society's Rumford Medal for his invention of the calotype process, Grove remarked that 'Mr. Talbot had a perfect right to patent his invention, but has on that account no claim in respect of the same invention to an honorary reward.'[9] So in the Court of Common Pleas (where he would eventually receive his first appointment as a judge) in December 1854 Grove found himself in the slightly invidious position of trying to defend Fox Talbot's patent against infringement when he did not believe that the patent ought to have been taken out in the first place.

Grove's interventions were taking place in the context of debates about the usefulness of patents that had been rumbling on for half a century or more. Patents were meant to protect inventors from having their inventions exploited by others before they had had the opportunity to profit from them themselves. As such, they awarded the patent holder the right of exclusive use of their invention for a period of seven years. By the early years of the nineteenth century the patent system was under attack on a number of fronts. Some claimed they were too expensive and favoured wealthy manufacturers rather than artisan inventors. According to some commentators they were often too restrictive. They prevented progress by making it difficult for rival inventors to develop and improve another's invention. Others suggested that, on the contrary, they were far too easy to evade and did not give inventors enough protection against infringement.[10] Grove would have been well aware that electrical inventors were particularly prone to contention. Charles Wheatstone and William Fothergill Cooke, who jointly held the first English patent for an electromagnetic telegraph, could not even agree with each other which of them should properly be called its inventor. Proposals for patent reform ranged from calls for outright abolition to suggestions for extending their lifetime. The Patent Law Amendment Act passed in 1852 did not do much to end the argument.

Whilst the court upheld the validity of Fox Talbot's patent, it rejected the charge that Laroche had infringed it. Fox Talbot now had to decide whether he should proceed with his efforts to extend the life of his calotype patent. Grove was clear that he should not. His barrister's opinion of patents was clearly known to Fox Talbot: 'I am aware of your strong opinions upon this subject', he wrote to him on one occasion.[11] Grove's views had by now developed beyond the stance that taking out a patent was not a fit thing for a man of science to do. Like many others during this period, he seems to have become increasingly suspicious of the patent system as a whole, even as he became increasingly involved in patent litigation, such as his dealings on Fox Talbot's behalf. What his experience seems to have taught him was that the problem with patents was not simply one of scientific propriety, but it was the public utility, or inutility, of the whole regime. He was making a name for himself as a legal authority on patents. In an article in the *Jurist* in 1860 Grove called for wholesale reform of the system and the establishment of a special patent court, that would be responsible not just for adjudicating disputes, but for awarding the patents themselves.

Patents, he thought, had outlived their purpose:

> when the patent law first grew into existence, inventive genius was rare; those who devoted themselves to a life of thought and experiment for practical purposes were few and far between; they had time, without the competitive interference of others, to develope and perfect their inventions.

Now, however, 'inventors are so numerous, the progress of physical science has made such vast strides, that it is, at all events with regard to a great number of inventions, a question only of weeks or months when an invention is to be made'.[12] Patents were now too common, and both too easy to apply illegitimately and to evade. Presumably as a result of his developing status as a patent authority, Grove was invited in 1864 to become one of the royal commissioners appointed to inquire into the operation of the law of patents. It seems clear from the final report that Grove's views had strongly influenced the outcome. The commissioners did not go the whole way in recommending the establishment of a special patent court, but they did recommend that the judge in patent cases should be advised by qualified scientific assessors, 'taking as a general model of its constitution Your Majesty's High Court of Admiralty, assisted by the Trinity Brethren'.[13]

What Grove envisioned was that just as the Court of Admiralty was advised by experienced men chosen from the Trinity Brethren, judges dealing with patent cases would be advised by suitably qualified men of science, drawn from a similar pool of appropriate scientific authorities. The suggestion, had it been adopted, would have provided another way for men of science to demonstrate their usefulness to the state. Grove, it seems, was still looking for ways to bring the two strands of his professional life together. He was looking for a role as a scientific lawyer. More than this, though, he was casting around for an appropriate public role for science too. One of the key elements in his agitation for scientific reform from the 1840s onwards was that if men of science wanted to find a proper public place for themselves they needed to reform their institutions to demonstrate their fitness for such offices. That was one of the roles he had wanted a properly reformed Royal Society to fulfil – making sure that the community of science presented a suitably disciplined face to the public and to potential paymasters in the corridors of power. Reforming the patent laws seemed an ideal avocation for a scientific lawyer, clearly, and offered another opportunity for Grove to make himself useful to science. If he could not have a scientific career, conventionally understood, he could still try to carve out another sort of space for his ambitions.

President of the BAAS

On the evening of Wednesday, 22 August 1866, Grove stepped onto the stage of Nottingham's 'large and handsome new theatre' to deliver his inaugural address as president of the BAAS. He would be performing in front of a packed house. According to the *Nottinghamshire Guardian*, the

> handsome and spacious building, which is capable of seating more than two thousand persons, exclusive of the unusually commodious stage and orchestra, was so crowded in every available nook and corner that many of the principal members of the association and their ladies were unable to gain admittance.

The stage itself was 'carpetted with crimson cloth, the president's desk, with the attached tables, being covered with green baize. At the sides and back of the stage, screens were arranged, giving to it the appearance of a fine hall.' It was crammed with dignitaries,

> including Lord Belper, the Lord Lieutenant of Nottinghamshire, Lord Wrottesley, Col. Sykes, M.P.; General Sabine, President of the Royal Society; Sir Roderick Murchison, Sir Charles Nicholson, Sir William Fairbairn, Sir John Bowring, Sir Edward Walker, High Sheriff of Nottinghamshire; Mr. Thomas Ball, Mayor of Nottingham; Mr. Hawksley, C.E.; Mr. Robert Crauford, President of the Ethnological Society; Professor Owen, Mr. James Glaisher, Mr. E. J. Lowe, and Mr. R. Birkin, jun.[14]

Grove was in some ways a slightly odd choice for the association's presidency on this occasion. Just like the marquis of Northampton when he graced the stage at the Swansea BAAS eighteen years earlier, Grove had no particular connection with the host town. Nottingham did not seem to mind, however, and in other respects Grove fitted the bill nicely. He still retained his reputation as an eminent man of science whilst also being increasingly known as a substantial public figure as a man of law. The new president was introduced to his audience by his predecessor, the geologist John Phillips, as a man who had 'long since declared those correlations among the forces of nature which allow of all being measured and weighed by one theoretic standard, and combined and directed in one great practical effort, honourable to a nation and beneficial to mankind'.[15] The BAAS's presidential addresses were traditionally quite conventional affairs. The incoming president looked back over the previous year, noting its scientific achievements and looking forward to the future. On this occasion, the audience may have been forewarned that Grove was planning to offer something a little different, and that there might even be a hint of controversy to the evening.

Grove began conventionally enough. He was not as he said, unaccustomed to public speaking, 'but on the contrary, I am perfectly accustomed to it'. What made this occasion different was that he was not 'accustomed to writing for the public, and the speech of the President of the British Association ought not to be spoken only – it ought to be read, and unfortunately for its author it must also be criticised'.[16] It was twenty years since he had done such a thing, he implied, and astute listeners might have grasped that this was the period of time that had elapsed since the publication of *Correlation*. It might have been understood as a hint of the direction the speech was to go. Grove was not, like some past presidents, going to offer 'a mere résumé of what has taken place since our last Meeting', he was going to talk about continuity, or the way in which

> the development of observational, experimental, and even deductive knowledge is either attained by steps so extremely small as to form really a continuous ascent; or, distinct results apparently separate from any coordinate phenomena have been attained, that then by the subsequent progress of science, intermediate links have been discovered uniting the apparently segregated instances with other more familiar phenomena.[17]

In other words, Grove was using the occasion to revisit some of the key themes of correlation, and the view that 'the more we investigate, the more we find that in existing phenomena graduation from the like to the seemingly unlike prevails, and in the changes which take place in time, gradual progress is, and apparently must be, the course of nature'.[18]

Grove began in the furthest reaches of heaven and proceeded to work his way down towards mere humanity. He offered a survey of recent years' discoveries in astronomy, discussing, among other things, Olmstead's work on meteorites and the implication from it that the 'universe would thus appear not to have the extent of empty space formerly attributed to it, but to be studded between the larger and more visible masses with smaller planets, if the term be permitted to be applied to meteorites'. He talked about the extension of his own work on the correlation of forces to the cosmos and speculated if 'our sun, our earth, and planets are constantly radiating heat into space, so in all probability are the other suns, the stars, and their attendant planets. What becomes of the heat thus radiated into space?' Since if

> the universe have no limit, and it is difficult to conceive one, heat and light should be everywhere uniform; and yet more is given off than is received by each cosmical body, for otherwise night would be as light and as warm as day. What becomes of the enormous force thus apparently non-recurrent in the same form?[19]

He talked about developments on geology and what they suggested about continuity.

Some of this may have reminded his older auditors of Robert Chambers's notorious *Vestiges of the Natural History of Creation*, written just a couple of years before Grove's original essay on *Correlation* twenty years previously, though coming from a far more impeccable authority. If so, they may have been primed for what came next. Grove argued that if his doctrine of continuity were to be taken seriously in discussions of the physical structure of the heavens and the earth, it should be taken just as seriously in discussions of the origins and development of organic life too. That, he suggested, was just what Darwin's theory of natural selection offered. 'I know I am touching on delicate ground,' he confessed, but I trust that the members of this body are sufficiently free from prejudice, whatever their opinions may be, to admit an inquiry into the general question whether what we term species are and have been rigidly limited, and have at numerous periods been created complete and unchangeable, or whether, in some mode or other, they have not gradually and indefinitely varied, and whether the changes due to the influence of surrounding circumstances, to efforts to accommodate themselves to surrounding changes, to what is called natural selection, or to the necessity of yielding to superior force in the struggle for existence, as maintained by our illustrious countryman Darwin, have not so modified organisms as to enable them to exist under changed conditions.[20]

If Darwin was right, then continuity should apply to the physical and cultural history of humanity as well, Grove suggested. The latest discoveries showed how

Man existed on this planet at an epoch far anterior to that commonly assigned to him. The instruments connected with human remains, and indisputably the work of human hands, show that to these remote periods the term civilization could hardly be applied – chipped flints of the rudest construction, probably, in the earlier cases, fabricated by holding an amorphous flint in the hand and chipping off portions of it by striking it against a larger stone or rock; then, as time suggested improvements, it would be more carefully shaped, and another stone used as a tool; then (at what interval we can hardly guess) it would be ground, then roughly polished, and so on, – subsequently bronze weapons, and, nearly the last before we come to historical periods, iron.

If so, then

what we call civilization must have been a gradual process; can it be supposed that the inhabitants of Central America or of Egypt suddenly and what is called instinctively built their cities, carved and ornamented their monuments? if not, if they must have learned to construct such erections, did it not take time to acquire such learning, to invent tools as occasion required, contrivances to raise weights, rules or laws by which men acted in concert to effect the design? Did not all this require time? and if, as the evidence of historical times shows, invention marches with a geometrical progression, how slow must have been the earlier steps![21]

Despite the controversial content the speech went down well. The 'admirable address of the President was deservedly cheered at its conclusion with enthusiastic demonstrations of approval; and a vote of thanks to Mr. Grove, on the motion of Lord Belper, Lord-Lieutenant of Nottinghamshire; seconded by Thomas Ball, Esq., Mayor of Nottingham, was carried by acclamation'.[22] Commentators further afield were not so sure. *The Times* thought Grove had overstepped the mark with his approval of Darwin's heresy. That was the point where he made the 'transition from the experimental field of science to the speculative'.[23] They took exception to his claim that it was easier to conceive an organism developing out of a previous one rather than appearing as if by a miracle. The *Pall Mall Gazette* came to Grove's defence, suggesting that *The Times* had missed the point. It was of course perfectly easy to conceive of an elephant appearing from nowhere, but that was not what Grove meant. He meant that 'such an event was improbable, as being altogether unlike our experience in other matters', and in this sense ease of conception was 'our only possible test for deciding whether it is probable or not'.[24] The generally supportive *Daily News* thought that by venturing into politics, Grove's speech had gone 'beyond the range of topics comprised in the scope of the Association' and thought his view of political philosophy 'hasty and ill-considered'.[25]

As for Darwin himself, he was in two minds about Grove's performance. He wrote to his friend Joseph Dalton Hooker that 'as a whole it strikes me as very good & original but I was disappointed in the part about species; it dealt in such generalities, that it would apply to any view or no view in particular'.[26] Hooker wrote back that Grove had deliberately left the detail for him to deal with:

I was to 'back him up' & 'to carry Darwinism through the ranks of the enemy' after he had sounded the charge: & whether or no his 'Continuity' Address was well received. In short I was a stink-pot, which he was to pitch into the Enemies decks, whether sinking or swimming himself.[27]

Grove himself responded that he had been

> anxious to put forward such arguments as seemed to me unanswerable & which addressed themselves not merely to specialists. I said not much about the adaptation view though entirely agreeing with it because the answer would have been that the argument cut both ways as whether an animal by circumstance Natural selection &c became suited to locality in conformation habits &c – or was specially created for particular circumstances the adaptation would equally be a necessity – an animal or plant must within limits be adapted to circumstance or not *be* at all.[28]

It seems likely that Grove's support for Darwin's theory was planned in advance and known to at least some others of Darwin's supporters – in fact Hooker later told Darwin that Grove had asked for his advice on the best available evidence to support Darwin's views.[29] It offers an intriguing insight into the extent to which Grove was still active in scientific circles and of the form his activity took. He was the established man offering support to a younger generation (though Darwin himself, of course, was his contemporary). He saw himself, maybe, as a scientific statesman. Hooker, for one, was clearly pleased to have his backing, telling Darwin that he

> felt flattered at the selection – puzzled as I was then, & am now, to make out why he should have thought me worthy of so responsible a post – on so critical an occasion. I had always a notion that he looked on me as a very weak vessel, & my branches of Botany as mild child's play.[30]

Grove's presidential address was a deliberate intervention that showed Grove off as a man willing to defend dangerous ideas, but to make those dangerous ideas part of the bigger picture of correlation that he had painted. In a way, it was a clever piece of scientific imperialism.

Final Years

Some of what the *Daily News* had to say about Grove offers a revealing insight into his reputation as his scientific name faded and his career in law continued to flourish. 'His scientific career,' they said, 'is a paper as instructive as any that have ever been submitted to the Association, and forms an unwritten Inaugural Address of the rarest value.' He was 'an example of the highest scientific eminence, reached by the intermitted labours of a man engaged day by day in the severest exertions of the most physically exhausting and mentally taxing of professions'. His achievements were all the more remarkable because the 'lawyer could give little help to the physicist, and the mental habits which the practice of advocacy tends to foster are ordinarily little favourable to the single-minded passion for truth which is the essential requisite of scientific investigation and its most valuable moral discipline'. Even more worthy of celebration was that his scientific fame lay not just in the discovery of 'mere fragments of truth', but in a theory that had transformed 'our entire conceptions of the material universe, establishing unity out of diversity, and detecting identity of force under the most dissimilar manifestations'.[31]

There was a tacit recognition in the encomium that Grove's great scientific days now lay behind him. His claim to fame as a natural philosopher lay with the *Correlation of Physical Forces*, a book that was now twenty years old, though a fourth edition had been published four years earlier, and a fifth edition would appear the following year, which included a new edition of his presidential address on continuity. The preface to the fifth editions offers some insight into where Grove himself thought he now stood in science. 'I have been regarded by many rather as the historian of the progress made in this branch of thought than as one who had anything to do with its initiation,' he complained.

He was quite clear that he was the 'first who introduced this subject as a generalised system of philosophy, and continued to enforce it in my lectures and writings for many years during which it met with the opposition usual and proper to novel ideas'.[32] The sixth edition of *Correlation*, published in 1874, also included a selection of his own scientific papers, titled 'Experimental Investigations', in order 'to avoid confusion with the better title appropriated by Faraday, of "Experimental Researches"'.[33]

After 1866, Grove only very rarely appeared in the public eye in the guise of a man of science. In 1869 he delivered an address to medical students on the importance of the study of the physical sciences for medical education at St Mary's Hospital in London. There he mounted another defence of correlation, arguing that it extended to the workings of the human body too. The electric battery was 'the nearest approach man has made to an artificial organism'.[34] Correlation was the reason medics should care about physics, since it taught them how the body fitted into the wider landscape of science. In 1872 he was invited to give evidence before the Devonshire Commission on Scientific and Technical Education. He used the opportunity to revisit the case for something like juxtaposition, calling for a reorganisation of scientific institutions that 'would save much useless expense, lead to more cooperation and promote more discipline in scientific workers, so that work in common for definite purpose might take the place of disjointed efforts'.[35] In 1888 he gave a Friday Evening Discourse at the Royal Institution on the topic of antagonism, in which he revisited again some of the themes of his address on continuity. Without antagonism 'there would be no light, heat, electricity, or life', he argued. Life without antagonism would be 'no life at all, a barren metaphysical conception of existence'.[36]

In 1888 Grove may well have been feeling the absence of antagonism, since, as he reminded his Royal Institution audience, he had only just relinquished his position as a high court judge. Where his scientific life had diminished, his legal career since the 1850s had flourished. In 1871 he was appointed as a judge of the Court of Common Pleas, which two years later became the Common Pleas Division of the High Court. In 1880 he was appointed to the Queen's Bench as one of the country's most senior judges. He had been knighted in 1872, shortly after his elevation to the bench. According to some accounts, one of the reasons he was made a judge was the assumption that his scientific knowledge would prove useful in trying patent cases, which in cases not involving the Crown would have been within the remit of the Court of Common Pleas. However, as one obituarist noted, 'he tried no more patent cases than other judges, and did not always try them well. The proper place for a scientific expert is the witness-box, and not the bench.' Grove had a reputation as a 'patient, courteous, laborious, and conscientious judge'.[37] Following his retirement in 1887 he was made a member of the Privy Council.

FIGURE 7 Cartoon of Mr Justice Grove, from *Vanity Fair*, 1887.

Grove died on 1 August 1896, aged eighty-five at his home in 115 Harley Street in London. He had been ill for some time. His wife, Emma, had died in 1879. Of his two sons and four daughters, two daughters had also predeceased him. Despite the fact that he had spent the last three active decades of his life dedicated to the law, his obituarists were almost unanimous in agreeing that his real eminence lay in science. *The Times* (which actually devoted more attention to his legal career than most other papers) even speculated that had Grove been able to study natural philosophy at Oxford instead of the usual classical education, then 'the whole and not a mere fragment of the future Judge's intellectual powers might have been devoted to physical studies – in which case it may be fairly conjectured that science would have gained more than law would have lost'.[38] The *Daily News* noted that it was 'not as a Judge that he will be chiefly remembered. He was one of the most distinguished men of science in a country and a time which produced Darwin and Faraday, Huxley and Tyndall, Sir John Herschel and Sir Charles Lyell.' To see him 'trying a running-down case, or an action of slander was perhaps as conspicuous an instance of misapplied talent and wasted energy as even the annals of the law could furnish'.[39]

Welsh newspapers recorded his death as well. Obituarists and correspondents made much of his local connections. The *Cambrian* detailed his Swansea upbringing and recalled how

> Swansea watched his distinguished career with keen and sympathetic interest, and applauded each step be took in his life's work ... He was little known to the younger generation of Swanseaites, except in name. Those who remember him when he studied and worked here, and came in contact with him upon his rare visits to his native town in after years, speak of him as being very reserved and somewhat irritable in disposition.

The obituarist also noted that he

> possessed considerable property in Carmarthen, Swansea, and Mumbles, chiefly in ground rents, which he seemed to regard

as the safest of investments. He owned all the ground rents in Henrietta-street, and also many in Bellevue-street, *Grove*-place, Trinity-place, Mount Pleasant, Worcester-place, Castle-street, Castle-square, High-street, Prince of Wales-road, &c.[40]

There were suggestions from others, even as proposals for a memorial bust at the Royal Institution of South Wales were canvassed, that 'he might have done more for his native town'.[41] Cardiff's *Evening Express* was more forthright. They recorded how during his second stint as president of the Royal Institution of South Wales in 1887–8 'a valuable address was expected from him, but he did not even condescend to visit the institution during the period of his presidency'.[42]

AFTERWORD

So just who was William Robert Grove when he died at his Harley Street residence in the summer of 1896? He was buried in London on 4 August. The funeral

> took place on Tuesday morning quite quietly at Kensal-green, The mourners included General Grove, Mr. Craufurd Grove [his two sons], Captain Duff Baker [husband of one of Grove's granddaughters], Mr. M. Crackenthorpe, Q.C., Mr. Herbert Crackenthorpe [father of one of Grove's sons-in-law and the son-in-law, respectively], Colonel I. Benson, Mr. John Hills, and Mr. Herbert Hills [son-in-law and grandson]. Wreaths and crosses of flowers were sent by several friends, but the deceased's known desire to have the funeral private was respected.[1]

He also died an extremely wealthy man: his estate amounted to a total of £215,899 14s. 4d, of which the bulk was inherited by his sons with substantial sums to other family members. The Swansea properties inherited from his father, along with his own investments and (presumably) the inheritance from his father-in-law John Diston Powles, had clearly done very nicely.[2] The only scientific item mentioned was his six foot telescope, left to his grandson Edmond Herbert Hills, himself an accomplished astronomer and fellow of the Royal Society.

To the casual eye, this looks like the will of a very successful late Victorian gentleman of independent means. There is not much here to remind us of his reputation as a man of science or even of his career in the law. It is clear from his obituaries that his scientific contributions still counted more to his contemporaries than his more recent achievements as a barrister and judge. Grove was the author of the *Correlation of Physical Forces*, the inventor of an important new kind of battery (though not the one he is remembered for now) and a leading light in the reform of the Royal Society during the 1840s. His obituarists agreed, though, that by the end of his life those achievements were a long way behind him. Predictably, only Welsh memorialists made anything at all of Grove's Welshness. The Swansea and south Wales papers made much of his status as a native son made good, though they too acknowledged that it had been some time since he had anything to do with Swansea other than as a source of income from property and rents. The one thing that makes Grove a famous Welshman now was entirely absent from any accounts of his achievements at the time of his death.

That one thing is, of course, the gas battery, or the fuel cell as we now call it. It is Grove's status as the Victorian inventor of a cutting-edge twenty-first-century technology that makes him an important figure again after more than a century in history's footnotes. The fuel cell is now widely anticipated to play a key role in transforming energy consumption as a way of combating global climate change. With fuel cells powering our cars our transport systems would be running on hydrogen and emitting water, rather than consuming fossil fuels and generating greenhouse gases. So it is quite easy now, as I did at the beginning of this book, to imagine some alternative histories of the future in which Grove moves from the margins and onto centre stage in the history of technology. It is a salutary reminder of the ways in which contemporary technological change and innovation change the ways in which we understand past technologies. Grove now seems set to join the pantheon of 'fathers' of new technologies. He will be remembered as the 'father of the fuel cell' just as James Watt is celebrated as the 'father of the steam engine'. It is a transformation in his reputation that would not have made much sense at all to Grove himself and his contemporaries.

We can get a good sense of just what the gas battery meant to Grove and his fellow electricians by looking at the prominent place it occupied in the frontispiece of Henry Minchin Noad's *Lectures on*

Electricity.³ This picture is a veritable Aladdin's cave of early Victorian electrical apparatus. Batteries jostle with induction coils that compete for attention with magneto-electric machines, or a tub of electric eels. In the foreground was a hydro-electric machine – a device that generated huge quantities of electricity from high pressure steam. In the middle of it all was Grove's gas battery. That was where it belonged in the 1840s. It was an instrument made to make electricity visible in spectacular style to the audiences at Grove's London Institution lectures. For Grove too, it was a technology to put his doctrine of the correlation of physical forces on display as well. Certainly useful instruments could be part of the technology of display too – the new Cooke and Wheatstone telegraph can be seen in Noad's frontispiece – but as far as Grove was concerned the gas battery was primarily useful as an illustration of a universal natural law, not as an economic source of power. It was a 'beautiful instance of the correlation of natural forces'.[4]

Grove himself clearly thought that it was his doctrine of the correlation that should be regarded as the cornerstone of his reputation. It was clear even in the first edition that he thought that the claims he was making about the interrelationships between the various forces of nature and the nature of scientific knowledge to be important and groundbreaking. He reiterated that view in all subsequent editions, particularly as he came to feel that his contribution to establishing this important doctrine was not getting the recognition he thought it deserved. He was not just the historian or the populariser of a new physics, he was its originator. Correlation might have started out as a way of talking about the relationship between different experiments performed in the London Institution's lecture theatre, but over the next few decades it developed into a powerful overarching view of the unity of nature – and a way of talking about the unity of nature that was widely used for the rest of the century. The language of correlation certainly offered strong competition to the doctrine of the conservation of energy as developed by James Prescott Joule, William Thomson, James Clerk Maxwell and others. Conflations of the two ways of talking were common, much to the annoyance of the promoters of energy as the defining concept in physics.

Figure 8 Frontispiece of Henry M. Noad, *Lectures on Electricity*, 1844.

Grove himself was happy to promote the idea that the conservation of energy was simply an extension (though couched in unnecessarily abstruse mathematical language) of his own original work on correlation. Unsurprisingly, men like Thomson and Maxwell, who regarded the theory of the conservation of energy as expressed in their work as the foundation of an entirely new approach to physics, saw things rather differently. As far as they were concerned, Grove and his theory of correlation belonged to an older and outmoded tradition that their new physics of energy was set to replace.[5] As we have seen, Peter Guthrie Tait, co-author with William Thomson of the *Treatise on Natural Philosophy*, denounced correlation as 'humbug'. Nevertheless, in his *The Unseen Universe*, co-authored with Balfour Stewart, he still acknowledged that 'Sir W. R. Grove did good work ... in bringing together the various cases of such transmutations in his work on the Correlation of the Physical Forces.'[6] Maxwell, when he reviewed the last edition of *Correlation* was just as anxious to draw a clear line in the sand between Grove's old science and his new one.

Grove, as we have seen, thought of correlation as far more than just a theory, however groundbreaking. It was a blueprint for how science in Britain ought to organise itself too. Grove's contemporaries certainly understood the role he had played in reforming the Royal Society during the 1840s, whether or not they agreed with those changes. For his allies, Grove had been a key figure in the process of transforming the Royal Society into an institution that was fit once more to take charge of the organisation of British science. Limiting the number of fellows elected each year and bringing the process under the control of the Royal Society's council was a way of bringing discipline back to its proceedings. Once disciplined, the society could resume its leading role at the head of science. Grove was trying to carve out a space for himself within the institutions of metropolitan science. His opponents saw things rather differently. From their perspective this was an attempt to wrest control of the Royal Society from the fellowship and centralise power in the hands of a small clique of self-styled scientific oligarchs who wanted to act as a 'higher council' over their fellows. The reformers were leading a scientific coup.

Historians of the Royal Society and of the professionalisation of science in the nineteenth century have cast the reforms spearheaded by Grove as a vital step in the process of turning the society into a properly professional body of professional scientists.[7] From this point of view, the move to limit the number of fellows elected each year and to restrict the fellowship to those who were 'qualified to fulfil the objects of the institution of the Royal Society', was designed to ensure that only properly professional scientists could be fellows. That is not how the reforms would have been understood at the time. There were very few men of science at the beginning of the Victorian period who could be described as professional – Grove until he resigned his position at the London Institution was one of them. No one took the view that only such individuals were qualified to be fellows of the Royal Society and few, if any, would have agreed that turning science into a profession was either necessary or desirable. Grove and his allies regarded science as a vocation for disciplined gentlemen. That is probably how Grove saw himself. He was a gentleman of science and his science was a vocation, not a profession. It is probably how most of his contemporaries regarded him too.

Grove's Swansea roots certainly played a role in his making as a Victorian gentleman of science. The connections he forged in south Wales with people like Lewis Weston Dillwyn who were already established figures on the national scientific stage, or with families like the Talbots, were to play a crucial role in his own launching onto the metropolitan scientific scene. Whilst Grove was coming of age as a man of science, Wales was in the process of developing its own scientific culture too. Scientific societies proliferated and Grove's own involvement with the Royal Institution of South Wales would be an important factor in his development as a practitioner rather than simply a spectator of science. The place provided him with an opportunity to hone his skills as a public performer. It seems clear that throughout the 1830s and much of the 1840s Grove moved back and forth between London and Swansea. Even as he was becoming an established figure in metropolitan science as the London Institution's professor of experimental philosophy, a fellow and then a council member of the Royal Society, the Swansea connection still mattered to him. It is just as clear, though, if his obituarists in the Welsh newspaper are to be believed, that by the end of his life Swansea mattered to him very little indeed.

Whilst Grove was clearly a man of science of Wales (he would not have called himself a scientist), was he a Welshman of science? There were certainly points in his career when a Welsh identity mattered. The British Association for the Advancement of Science's (BAAS) Swansea visit is a good example. Persuading both the BAAS's leadership and potential supporters in Wales that this was a specifically Welsh occasion was an important part of Grove's strategy for attracting the BAAS to come to town. This was to be a visit to the 'metropolis of Wales'. Northampton in his address was just as keen to emphasise Grove's Welsh identity. In his case it was a way of putting his chief tormentor in his place. Whether it makes sense for us now to describe Grove as a specifically Welsh man of science is clearly an open question. It was an identity that Grove adopted when it suited, just as others could label him in that way for their own reasons. What matters, both for understanding Grove and for understanding the place of science in nineteenth-century Wales, is that we can see how Wales and Swansea's scientific culture contributed to making him what he was.

NOTES

Prologue

1. Carlo Matteucci, 'On the Electricity of Flame, with Comments by Mr. Grove', *Philosophical Magazine*, 8 (1854), 399–404, on 403–4.
2. Gregor Hoogers, *Fuel Cell Technology Handbook* (Boca Raton: CRC Press, 2002).
3. Charles Kingsley, *The Water Babies* (London: Macmillan, 1863), p. 66.
4. J. G. Crowther, *Statesmen of Science* (London: Cresset Press, 1965), pp. 77–101.
5. Quoted in Crosbie Smith, *The Science of Energy: A Cultural History of Energy Physics in Victorian Britain* (London: Athlone Press, 1998), p. 176.
6. William Robert Grove, 'On the Progress made in the Application of Electricity as a Motive Power', *Literary Gazette*, 28 (1844), 113.
7. William Robert Grove, *The Correlation of Physical Forces*, 6th edn (London: Longmans, Green and Co., 1874), p. 140.
8. William Robert Grove, 'Presidential Address', *Report of the British Association for the Advancement of Science*, 26 (1866), pp. liii–lxxxi, on lxxix.
9. Grove, 'Presidential Address', p. lxxix.
10. *Welsh Heroes* (Aberystwyth: Culturenet Cymru, 2004).

Chapter 1

1. Ronald Rees, *King Copper: South Wales and the Copper Trade, 1584–1895* (Cardiff: University of Wales Press, 2000), p. 72. Rees wrongly identifies this Grove with his younger scientific namesake. William Robert Grove the younger would need to have been remarkably precocious to have been involved in such activities at the age of eleven.
2. I am extremely grateful to Gerald Gabb for much of the information regarding the Grove family contained in the preceding two paragraphs.
3. 'The Late Sir W. Grove: Interesting Letter', *Cambrian*, 14 August 1896.
4. 'The Latest Finds on the Castle Bailey Site', *Cambrian*, 31 March 1893.
5. 'Swansea Grammar School', *Welshman*, 10 September 1841.
6. 'William Robert Grove', *Portraits of Men of Eminence in Literature, Science, and Art, with Biographical Memoirs, the Photographs from Life, by Ernest Edwards, BA* (London: Alfred William Bennett, 1865), vol. 3, pp. 29–30.
7. Francis Galton, *English Men of Science: Their Nature and Nurture* (London:

Macmillan, 1874), p. 154. The section on Grove in this exposition of Galton's hereditary theory is identifiable by its similarities to the biographical details offered in *Portraits*.
8. 'Death of Sir William Grove', *Cambrian*, 7 August 1896.
9. John Toman, *Kilvert's World of Wonders: Growing Up in Mid-Victorian England* (Cambridge: Lutterworth Press, 2013).
10. *Portraits*, pp. 29–30.
11. Louise Miskell, *Intelligent Town: An Urban History of Swansea, 1780–1855* (Cardiff: University of Wales Press, 2006).
12. Quoted in Miskell, *Intelligent Town*, p. 50.
13. Quoted in Miskell, *Intelligent Town*, p. 53.
14. *Merthyr Guardian*, 9 May 1835.
15. *Cambrian*, 2 May 1835.
16. *Cambrian*, 5 September 1835.
17. 'Professor Airy's Lecture at Neath', *Cambrian*, 16 September 1837.
18. 'Literary and Scientific Institution at Merthyr', *Cambrian*, 14 October 1837.
19. 'Swansea Philosophical Institution', *Cambrian*, 19 December 1835.
20. W. A. Beanland, *The History of the Royal Institution of South Wales* (Swansea: Royal Institution of South Wales, 1935), quote on p. 20.
21. Pietro Corsi, *Science and Religion: Baden Powell and the Anglican Debate, 1800–1860* (Cambridge: Cambridge University Press, 1988). Also Robert Fox and Graeme Gooday (eds), *Physics in Oxford, 1839–1939* (Oxford: Oxford University Press, 2005).
22. For the early history of the BAAS, see Jack Morrell and Arnold Thackray, *Gentlemen of Science: Early Years of the British Association for the Advancement of Science* (Oxford: Oxford University Press, 1981).
23. 'Swansea Literary and Philosophical Institution', *Cambrian*, 28 January 1837.
24. 'Swansea Literary and Philosophical Institution', *Cambrian*, 28 January 1837.
25. *Cambrian*, 12 January 1839.
26. 'British Association for the Advancement of Science', *Cambrian*, 7 September 1839.

Chapter 2

1. One example is 'William Robert Grove', *Portraits of Men of Eminence in Literature, Science, and Art, with Biographical Memoirs, the Photographs from Life, by Ernest Edwards, BA* (London: Alfred William Bennett, 1865), vol. 3, pp. 29–30. Some obituaries (such as 'Death of Sir William Grove', *Cambrian*, 7 August 1896) also mention poor health as a factor.
2. For an overview, see Iwan Rhys Morus, Simon Schaffer and James Secord, 'Scientific London', in Celina Fox (ed.), *London: World City, 1800–1840* (London and New Haven: Yale University Press, 1992), pp. 129–42.
3. Carlyle's comment is in a letter to his brother Alexander, 5 March 1837, in Alexander Carlyle (ed.), *New Letters of Thomas Carlyle* (London: Bodley Head, 1904), vol. 1, pp. 62–3. For the early history of the Royal Institution, see Morris Berman, *Social Change and Scientific Organization: The Royal Institution, 1799–1844* (London:

Heinemann, 1978).
4. George Eliot, *The Mill on the Floss*, Great Writers edn (1860; London: Marshall Cavendish, 1986), p. 289.
5. Harold Silver, *English Education and the Radicals 1780–1850* (London: Routledge and Kegan Paul, 1977).
6. J. N. Hays, 'The London Lecturing Empire, 1800–50', in Ian Inkster and Jack Morrell (eds), *Metropolis and Province: Science in British Culture 1780–1850* (London: Hutchison, 1983), pp. 91–119.
7. Iwan Rhys Morus, 'Worlds of Wonder: Sensation and the Victorian Scientific Performance', *Isis*, 101 (2010), 806–16.
8. Richard Altick, *The Shows of London* (Cambridge, MA: Belknap Press, 1978) offers a survey of these places.
9. Albert Smith, *Gavarni in London: Sketches of Life and Character* (London: David Bogue, 1849), p. 13.
10. Edmund Yates, *Recollections and Experiences*, 2 vols (London: Richard Bentley & Son, 1884), vol. 1, p. 136.
11. Iwan Rhys Morus, *Frankenstein's Children: Electricity, Exhibition and Experiment in Early Nineteenth-century London* (Chicago: University of Chicago Press, 1998), pp. 83–6.
12. Altick, *The Shows of London*.
13. Morus, *Frankenstein's Children*, pp. 83–6.
14. William Sturgeon, 'Account of Improved Electro-magnetic Apparatus', *Annals of Philosophy*, 12 (1826), 357–61.
15. Iwan Rhys Morus, 'Different Experimental Lives: Michael Faraday and William Sturgeon', *History of Science*, 30 (1992), 1–28.
16. *Cambrian*, 3 June 1837.
17. I thank Frank James for this information.
18. William Robert Grove, 'On a New Voltaic Combination', *Philosophical Magazine*, 13 (1838), 430–1, on 430–1.
19. Grove, 'On a New Voltaic Combination', 431.
20. William Robert Grove, 'On Voltaic Series and the Combination of Gases by Platinum', *Philosophical Magazine*, 14 (1839), 127–30, on 128.
21. William Robert Grove, 'On a New Voltaic Battery, and on Voltaic Combinations and Arrangements', *Philosophical Magazine*, 14 (1839), 287–93, on 287.
22. 'Royal Institution', *Literary Gazette*, 24 (1840), 184.
23. *Historical Account of the London Institution* (London: London Institution, 1835), pp. 25–6.
24. Iwan Rhys Morus, 'Currents from the Underworld: Electricity and the Technology of Display in early Victorian London', *Isis*, 84 (1993), 50–69.
25. J. N. Hays, 'Science in the City: The London Institution, 1819–1840', *British Journal of the History of Science*, 7 (1974), 46–62; Berman, *Social Change and Scientific Organization*.
26. Charles Butler, *The Inaugural Oration, Spoken on the 4th Day of November 1815, at the Ceremony of Laying the First Stone of the London Institution for the Diffusion of Science*

and Literature (London: Longman, Hurst, Rees, Orme and Brown, 1816), p. 34.
27. Butler, *The Inaugural Oration*, p. 40.
28. William Robert Grove, *On the Progress of Physical Science since the Opening of the London Institution* (London: London Institution, 1842), p. 6.
29. Grove, *On the Progress of Physical Science since the Opening of the London Institution*, p. 17.
30. Grove, *On the Progress of Physical Science since the Opening of the London Institution*, p. 24.
31. Grove, *On the Progress of Physical Science since the Opening of the London Institution*, p. 37.
32. *Historical Account of the London Institution* (1835), p. 26.
33. 'London Institution', *Literary Gazette*, 26 (1842), 67.
34. *Historical Account of the London Institution* (1835), p. 30 n.
35. M. L. Cooper and V. M. D. Hall, 'William Robert Grove and the London Institution, 1841–1845', *Annals of Science*, 39 (1982), 229–54.
36. Charles Vincent Walker, 'Notice on the Relative Powers of certain Diaphragm Voltaic Combinations, and on a New Form of the Negative Element in Voltaic Arrangements', *Proceedings of the London Electrical Society*, 2 (1841), 114–18.
37. Iwan Rhys Morus, 'The Sociology of Sparks: An Episode in the History and Meaning of Electricity', *Social Studies of Science*, 18 (1988), 387–417.
38. Grove, 'On Voltaic Series, and the Combination of Gases by Platinum', 130.
39. William Robert Grove to Michael Faraday, 22 October 1842, in L. Pearce Williams (ed.), *The Selected Correspondence of Michael Faraday*, 2 vols (Cambridge: Cambridge University Press, 1971), vol. 1, p. 402.
40. William Robert Grove, 'New Voltaic Battery – Gaseous Elements', *Literary Gazette*, 26 (1842), 833.
41. William Robert Grove, 'On Some Phenomena of the Voltaic Disruptive Discharge', *Philosophical Magazine*, 16 (1840), 478–82.
42. William Robert Grove, 'On the Application of Voltaic Ignition to Lighting Mines', *Philosophical Magazine*, 27 (1845), 442–6.
43. 'Explosions in Coal Mines', *Cambrian*, 24 July 1846.
44. William Robert Grove, 'On Photography', *Reports of the British Association for the Advancement of Science*, 14 (1844), 37.
45. William Robert Grove, 'On a Voltaic Process for Etching Daguerreotype Plates', *Proceedings of the London Electrical Society*, 2 (1843), 94–100, on 100.
46. William Robert Grove to Michael Faraday, 19 December 1842, *Selected Correspondence*, vol. 1, p. 408.
47. William Robert Grove, 'Remarks on a Letter of Prof. DANIELL, in the Philosophical Magazine, "On the Constant Voltaic Battery"', *Philosophical Magazine*, 21 (1842), 333–5.

Chapter 3

1. 'Electrical Soirée', *Literary Gazette*, 26 (1842), 295–6, on 296.

2. 'Electrical Soirée', *Literary Gazette*, 27 (1843), 352.
3. 'Royal Polytechnic Institution', *Morning Chronicle*, 16 September 1843.
4. W. J. Copleston, *Memoir of Edward Copleston, D.D., Bishop of Llandaff* (London: John W. Parker & Son, 1851), p. 169.
5. 'Davenport's Electro-Magnetic Engine', *Mechanics' Magazine*, 27 (1837), 404–5.
6. Alfred Smee, *Elements of Electro-metallurgy, or the Art of Working in Metals by the Galvanic Fluid* (London: Longman, Rees, Orme, Brown & Longman, 1841), p. 147.
7. William Robert Grove, 'Experiments on Voltaic Reaction', *Philosophical Magazine*, 23 (1843), 45–7, on 446.
8. William Robert Grove, 'On the Application of Voltaic Electricity to Lighting Mines', *Philosophical Magazine*, 27 (1845), 442–6.
9. William Robert Grove, 'On Certain Phenomena of Voltaic Ignition and the Decomposition of Water into its Constituent Gases by Heat', *Philosophical Transactions*, 137 (1847), 1–16, on 1.
10. James Secord, *Visions of Science: Books and Readers at the Dawn of the Victorian Age* (Oxford: Oxford University Press, 2014).
11. John Herschel, *Preliminary Discourse on the Study of Natural Philosophy* (London: Longman, Rees, Orme, Brown & Green, 1831), p. 4.
12. *Preliminary Discourse on the Study of Natural* Philosophy, p. 15.
13. *Preliminary Discourse on the Study of Natural Philosophy*, p. 222.
14. Mary Somerville, *On the Connexion of the Physical Sciences*, 2nd edn (London: John Murray, 1835), p. 4.
15. Somerville, *On the Connexion of the Physical Sciences*, p. 416.
16. William Whewell, *History of the Inductive Sciences*, 3 vols (London: John Parker, 1837), vol. 1, p. xiv.
17. Whewell, *History of the Inductive Sciences*, p. 13.
18. Charles Babbage, *The Ninth Bridgewater Treatise: A Fragment* (London: John Murray, 1837), p. 115.
19. James Secord, *Victorian Sensation: The Extraordinary Publication, Reception and Secret Authorship of Vestiges of the Natural History of Creation* (Chicago: University of Chicago Press, 2003).
20. Robert Chambers, *Vestiges of the Natural History of Creation and Other Evolutionary Writings*, ed. James Secord (1844; Chicago: University of Chicago Press, 1994), p. 163.
21. Thomas Simmons Mackintosh, *The Electrical Theory of the Universe* (Boston: Josiah P. Mendum, 1846), p. 371.
22. William Robert Grove, *On the Correlation of Physical Forces* (London: London Institution, 1846), p. iii.
23. Grove, *On the Correlation of Physical Forces*, p. 5.
24. Grove, *On the Correlation of Physical Forces*, p. 7.
25. Grove, *On the Correlation of Physical Forces*, p. 8.
26. William Robert Grove, *On the Progress of Physical Science since the Opening of the London Institution* (London: London Institution, 1842), p. 31.
27. Grove, *On the Progress of Physical Science since the Opening of the London Institution*,

p. 8.
28. Grove, *On the Progress of Physical Science since the Opening of the London Institution*, 9.
29. 'London Institution', *Philosophical Magazine*, 24 (1844), 76–8, on 77.
30. Grove, *On the Correlation of Physical Forces*, p. 28.
31. William Robert Grove, 'New Voltaic Battery – Gaseous Elements', *Literary Gazette*, 26 (1842), 833.
32. Grove, *On the Correlation of Physical Forces*, pp. 42–4.
33. Grove, *On the Correlation of Physical Forces*, p. 50.
34. Richard Olson, *Scottish Philosophy and British Physics, 1750–1880* (Princeton: Princeton University Press, 1975) discusses the wider impact of common sense philosophy on Victorian physics.
35. Thomas Brown, *An Inquiry into the Relation of Cause and Effect* (Edinburgh: Archibald Constable, 1806), p. 17.
36. Charles Babbage, *On the Economy of Machinery and Manufactures* (London: Charles Knight, 1832), p. 39.
37. Peter Barlow, *Treatise on the Manufactures and Machinery of Great Britain* (London: Baldwin & Cradock, 1831), p. 91.
38. Barlow, *Treatise on the Manufactures and Machinery of Great Britain*, p. 397.
39. Andrew Ure, *The Philosophy of Manufactures* (London: Charles Knight, 1835), p. 1.
40. Ure, *The Philosophy of Manufactures*, p. 14.
41. William Robert Grove, 'On the Progress Made in the Application of Electricity as a Motive Power', *Literary Gazette*, 28 (1844), 113.
42. 'On the Correlation of Physical Forces', *British and Foreign Medical Review*, 23 (1847), 247–8, on 248.
43. 'Grove, Carpenter &c. on the Correlation of Forces, Vital and Physical', *British and Foreign Medico-chirurgical Review*, 8 (1851), 206–37, on 207.
44. 'On the Correlation of Physical Forces', *British Quarterly Review*, 14 (1851), 155–78, on 156.
45. Grove, *On the Correlation of Physical Forces*, p. 49.
46. William Benjamin Carpenter, 'On the Mutual Relations of the Vital and Physical Forces', *Philosophical Transactions*, 140 (1850), 727–57.
47. 'Grove, Carpenter &c. on the Correlation of Forces, Vital and Physical', 226.
48. Thomas Laycock, *Mind and Brain, or, the Correlations of Consciousness and Organisation* (London: Simpkin, Marshall & Co., 1859).
49. George Eliot, *Middlemarch: A Study in Provincial Life*, 8 vols (London: William Blackwood & Sons, 1872), vol. 2, p. 321.
50. Edward Bulwer-Lytton, *The Coming Race* (1871; Stroud: Alan Sutton Publishing, 1995), p. 20.
51. Charles Kingsley, *The Water Babies* (London: Macmillan, 1863), p. 66.
52. Karl Marx to Friedrich Engels, 31 August 1864.
53. James Clerk Maxwell, 'Grove's Correlation of Physical Forces', *Nature*, 10 (1874), 302–4, on 303.
54. Maxwell, 'Grove's Correlation of Physical Forces', 303. For Maxwell and his followers, see Bruce Hunt, *The Maxwellians* (Ithaca: Cornell University Press, 1994).

Chapter 4

1. Marie Boas Hall, *Promoting Experimental Learning: Experiment and the Royal Society, 1660–1727* (Cambridge: Cambridge University Press, 1991), p. 9.
2. Larry Stewart, *The Rise of Public Science* (Cambridge: Cambridge University Press, 1992).
3. John Gascoigne, *Joseph Banks and the English Enlightenment* (Cambridge: Cambridge University Press, 1994).
4. Donald Cardwell, *The Organisation of Science in England* (London: Heinemann, 1972); David P. Miller, 'Method and the Micropolitics of Science: The Early Years of the Geological and Astronomical Societies', in John Schuster and Richard Yeo (eds), *The Politics and Rhetoric of Scientific Method* (Dordrecht: D. Reidel & Co., 1986), pp. 227–58.
5. David P. Miller, 'Between Hostile Camps: Sir Humphry Davy's Presidency of the Royal Society of London, 1824–27', *British Journal for the History of Science*, 16 (1983), 1–47.
6. Royal Society Library, Council Minutes 3 May 1827.
7. Charles Babbage, *Reflections on the Decline of Science in England* (London: B. Fellowes, 1830), p. 160.
8. Babbage, *Reflections on the Decline of Science in England*, p. 165.
9. Marie Boas Hall, *All Scientists Now* (Cambridge: Cambridge University Press, 1984).
10. Quoted in M. Jeanne Peterson, *The Medical Profession in mid-Victorian London* (Berkeley and Los Angeles: University of California Press, 1978), p. 26. See also Adrian Desmond, *The Politics of Evolution* (Chicago: University of Chicago Press, 1989).
11. Quoted in Harvey Becher, 'Radicals, Whigs and Conservatives: The Middle and Lower Classes in the Analytical Revolution in Cambridge in the Age of Aristocracy', *British Journal for the History of Science*, 28 (1995), 405–26, on 411.
12. L. Pearce Williams (ed.), *The Selected Correspondence of Michael Faraday*, 2 vols (Cambridge: Cambridge University Press, 1971), vol. 1, p. 408.
13. John Frederic Daniell, 'On the Constant Voltaic Battery', *Philosophical Magazine*, 20 (1842), 294–304, on 304.
14. William Robert Grove 'On the Subject of Mr. DANIELL's Last Communication', *Philosophical Magazine*, 22 (1843), 32–5, on 32.
15. Grove 'On the Subject of Mr. DANIELL's Last Communication', 35.
16. Michael Faraday to William Robert Grove, 22 December 1842, Grove Correspondence Royal Institution (henceforward GCRI).
17. [William Robert Grove], 'Physical Science in England', *Blackwood's Magazine*, 54 (1843), 514–25, on 517–18.
18. [Grove], 'Physical Science in England', 518.
19. Royal Society Council Minutes, 7 May 1846.
20. Katherine Lyell, *Memoir of Leonard Horner*, 2 vols (London: Women's Printing Society, 1890).
21. There is a copy of this letter, Leonard Horner to Peter Mark Roget, 9 May 1846, in GCRI.

22. Royal Society Library, Committee Minutes. Charter Committee Minutes are in DM.1 nos. 40–60 (henceforward DM.1), 11 May 1846.
23. Leonard Horner to William Robert Grove, 15 May 1846, GCRI.
24. Royal Society Council Minutes, 28 May 1846.
25. DM.1 1 June 1846; Royal Society Council Minutes, 4 June 1846.
26. Royal Society Council Minutes, 18 June 1846.
27. Royal Society Council Minutes, 18 June 1846.
28. Leonard Horner to William Robert Grove, 28 October 1846, GCRI.
29. Royal Society Council Minutes, 5 November 1846.
30. Leonard Horner to William Robert Grove, 10 November 1846, GCRI.
31. Leonard Horner to William Robert Grove, 11 November 1846, GCRI.
32. Babbage, *Reflections on the Decline of Science in England*, pp. 62–5.
33. DM.1 25 November 1846.
34. Leonard Horner to William Robert Grove, 24 November 1846, GCRI.
35. DM.1 25 November 1846. See also Henry Lyons, *The Royal Society 1660–1940: A History of its Administration under its Charters* (Cambridge: Cambridge University Press, 1944). Presumably the fragment was part of the report that Grove circulated to Horner and Sabine.
36. Royal Society Council Minutes 26 November 1846.
37. Leonard Horner to William Robert Grove, 9 December 1836, GCRI.
38. Royal Society Council Minutes 17 December 1846.
39. Royal Society Council Minutes 14 January 1847.
40. Royal Society Council Minutes 21 January 1847.
41. Leonard Horner to William Robert Grove, 18 January 1847, GCRI.
42. Henry Lyons, *Record of the Royal Society* (London: Royal Society, 1940), pp. 301–15.
43. Marquis of Northampton, 'Presidential Address', *Proceedings of the Royal Society*, 5 (1843–50), 698–703, on 703.
44. Marquis of Northampton, 'Presidential Address', *Proceedings of the Royal Society*, 5 (1843–50), 761–7, on 764.
45. Marquis of Northampton, 'Presidential Address', 762–73.
46. *Proceedings of the Royal Society*, 5 (1843–50), 767.
47. *Proceedings of the Royal Society*, 5 (1843–50), 827.

Chapter 5

1. Jack Morrell and Arnold Thackray, *Gentlemen of Science: Early Years of the British Association for the Advancement of Science* (Oxford: Oxford University Press, 1981).
2. A. D. Orange, *Philosophers and Provincials: The York Philosophical Society 1822–1844* (York: York Philosophical Society, 1973).
3. Charles Babbage, *Reflections on the Decline of Science in England* (London: B. Fellowes, 1830).
4. Sydney Ross, 'Scientist: The Story of a Word', *Annals of Science*, 19 (1962), 65–85.
5. Charles Dickens, *The Mudfog Papers* (London: Richard Bentley and Son, 1880). They

had first been published as a series of sketches in *Bentley's Miscellany*, then edited by Dickens, from 1837 to 1838.
6. Louise Miskell, *Intelligent Town: An Urban History of Swansea, 1780–1855* (Cardiff: University of Wales Press, 2006).
7. *Reports of the British Association for the Advancement of Science*, 16 (1847), xviii.
8. 'British Association', *Cambrian*, 12 September 1846.
9. 'British Association', *Cardiff and Merthyr Guardian*, 3 July 1847.
10. David Williams, *The Rebecca Riots* (Cardiff: University of Wales Press, 2011).
11. RISW/DL 49, 'Pocket book of John Dillwyn-Llewelyn …', Dillwyn-Llewelyn Collection, Royal Institution of South Wales Collection, West Glamorgan Archive Service, Swansea. My thanks to Jim Moore for drawing this reference to my attention.
12. 'Eighteenth Meeting of the British Association for the Advancement of Science', *Cardiff and Merthyr Guardian*, 8 July 1848.
13. 'British Association', *Cardiff and Merthyr Guardian*, 15 July 1848.
14. Lewis Weston Dillwyn Diaries, entries for 9–19 August, National Library of Wales.
15. D. S. M. Barrie, *Regional History of the Railways of Great Britain: South Wales* (London: David and Charles, 1980).
16. 'Swansea. – British Association', *Cardiff and Merthyr Guardian*, 12 August 1848.
17. 'British Association for the Advancement of Science', *Pembrokeshire Herald*, 18 August 1848.
18. Map of Swansea for visitors to the BAAS Meeting. Royal Institution of South Wales.
19. Marquis of Northampton, 'Presidential Address', *Report of the British Association for the Advancement of Science*, 18 (1848), pp. xxxi–xxxix, on p. xxxii.
20. Marquis of Northampton, 'Presidential Address', p. xxxi.
21. Marquis of Northampton, 'Presidential Address', p. xxxii
22. Marquis of Northampton, 'Presidential Address', p. xxxiii.
23. 'Eighteenth Meeting of the British Association for the Advancement of Science', *Athenaeum* (1848), 863–4, on 863.
24. Matthew Moggridge, 'On Two Cases of Uncommon Atmospheric Refraction', *Report of the British Association for the Advancement of Science*, 18 (1848), 33–4, on 34.
25. James Prescott Joule, 'On the Mechanical Equivalent of Heat, and the Constitution of Elastic Fluids', *Report of the British Association for the Advancement of Science*, 18 (1848), 21–2, on 23.
26. William Robert Grove, 'On the Peculiar Cooling Effects of Hydrogen and its Compounds in Cases of Voltaic Ignition', *Report of the British Association for the Advancement of Science*, 18 (1848), 55.
27. James Nasmyth, 'On a Peculiar Property of Coke', *Report of the British Association for the Advancement of Science*, 18 (1848), 56.
28. W. S. Ward, 'On a Galvanometer', *Report of the British Association for the Advancement of Science*, 18 (1848), 62.
29. Spence Bate, 'On Fossil Remains Recently Discovered in Bacon Hole, Gower; also Other Remains from Beneath the Bed of the River Tawey', *Report of the British Association for the Advancement of Science*, 18 (1848), 62–3.

30. Starling Benson, 'On the Relative Positions of the Various Qualities of Coal in the South Wales Coal-measures', *Report of the British Association for the Advancement of Science*, 18 (1848), 65–6.
31. Henry de la Beche, 'On the Geology of Portions of South Wales, Gloucestershire and Somersetshire', *Report of the British Association for the Advancement of Science*, 18 (1848), 79.
32. Lewis Weston Dillwyn Diaries, 10 August 1848.
33. Matthew Moggride, 'On a Peculiarity in the Protococcus Nivalis', *Report of the British Association for the Advancement of Science*, 18 (1848), 86.
34. Thomas Williams, 'On the Physical Conditions Regulating the Vertical Distribution of Animals in the Atmosphere and the Sea', *Report of the British Association for the Advancement of Science*, 18 (1848), 83.
35. Adrian Desmond, *The Politics of Evolution* (Chicago: University of Chicago Press, 1989); Samuel Wilks and George Bettany, *A Biographical Dictionary of Guy's Hospital* (London: Ward, Lock, Bowden and Co., 1892), p. 439.
36. Edwin Lankester, 'On Some Vegetable Monstrosities Illustrating the Laws of Morphology', *Report of the British Association for the Advancement of Science*, 18 (1848), 85–6.
37. Richard Owen, 'On the Communication between the Tympanum and the Palate in the Crocodiles', *Report of the British Association for the Advancement of Science*, 18 (1848), 79–80.
38. Professor Elton, 'On the Ante-Columbian Discovery of America', *Report of the British Association for the Advancement of Science*, 18 (1848), 95–6.
39. Cadogan Williams, 'On the Desirability of Extending to the Working Classes the opportunity of Purchasing Deferred Annuities, as a Provision for Old Age', *Report of the British Association for the Advancement of Science*, 18 (1848), 105.
40. Joseph Fletcher, 'Statistics of Brittany and the Bretons', *Report of the British Association for the Advancement of Science*, 18 (1848), 114.
41. J. Ashman, 'An Artificial Leg, of an Improved Construction', *Report of the British Association for the Advancement of Science*, 18 (1848), 117.
42. Francis Wishaw, 'On the Subaqueous Rope for Telegraphic and other Purposes', *Report of the British Association for the Advancement of Science*, 18 (1848), 123.
43. 'Meeting of the British Association at Swansea', *Welshman*, 18 August 1848.
44. 'Meeting of the British Association at Swansea'.
45. 'British Association – Swansea'. Pamphlet. Royal Institution of South Wales.
46. Robert Hunt, 'On Electro-magnetism as a Motive Power', *Proceedings of the Institution of Civil Engineers*, 16 (1857), 386–421. Grove made his claim regarding his role in the design of the engine in the discussion following the paper's presentation.
47. Benjamin Hill, 'On a New Electro-magnetic Machine', *Proceedings of the London Electrical Society*, 2 (1841), 83–6, on 83.
48. Lewis Weston Dillwyn Diaries, 19 August 1848. National Library of Wales.
49. 'Swansea – British Association', *Cardiff and Merthyr Guardian*, 25 August 1848.

Chapter 6

1. Leonard Horner to William Robert Grove, 25 February 1847, GCRI.
2. Leonard Horner to William Robert Grove, 25 February 1847, GCRI.
3. John Peter Gassiot to John Herschel, 2 March 1847. Herschel Papers, Royal Society Library.
4. Thomas Bonney, *Annals of the Philosophical Club of the Royal Society* (London: Macmillan, 1919), p. 1.
5. Bonney, *Annals of the Philosophical Club of the Royal Society*, pp. 1–3.
6. Leonard Horner to William Robert Grove, 10 October 1847.
7. Edward Forbes to William Robert Grove, n.d.
8. The two bound volumes of the Philosophical Club's Minute Book are in the Royal Society Library, MS.721.
9. Edward Forbes to William Robert Grove, 18 March 1848, GCRI.
10. Leonard Horner to William Robert Grove, 23 January 1848, GCRI.
11. Charles Mollan (ed.), *William Parsons, 3rd Earl of Rosse* (Manchester: Manchester University Press, 2016).
12. Royal Society Council Minutes, 13 April 1848.
13. Henry de la Beche to William Robert Grove, 25 April 1848, GCRI.
14. Katherine Lyell, *Life, Letters and Journals of Sir Charles Lyell*, 2 vols (London: John Murray, 1881), vol. 2, p. 145. This was another example of revolutionary rhetoric by the reformers in this year of revolutions. The ousted Austrian foreign minister Prince Klemens von Metternich had only recently fled Vienna for London when Lyell wrote this to Horner.
15. Charles Lyell to William Robert Grove, 7 July 1848, GCRI.
16. 'A JUNIOR F.R.S.', *Athenaeum*, 18 November 1848, 1149.
17. 'A PHYSICAL F.R.S.', *Athenaeum*, 25 November 1848, 1179.
18. 'Royal Society', *Literary Gazette*, 32 (1848), 408.
19. 'Royal Society', *Literary Gazette*, 32 (1848), 775.
20. 'Royal Society', *Lancet*, 2 (1848), 591.
21. William Bowman to William Robert Grove, 20 November 1848, GCRI.
22. William Benjamin Carpenter to William Robert Grove, 22 November 1848, GCRI.
23. 'Royal Society', *Literary Gazette*, 32 (1848), 792.
24. Leonard Horner to William Robert Grove, 1 December 1848, GCRI.
25. Charles Lyell to William Robert Grove, n.d. GCRI.
26. Iwan Rhys Morus, *When Physics Became King* (Chicago: University of Chicago Press, 2005), chapter 6.
27. William Robert Grove, 'On the Effect of Surrounding Media on Voltaic Ignition', *Philosophical Transactions*, 139 (1849), 49–59, on 49.
28. William Robert Grove, 'On the Electro-chemical Polarity of Gases', *Philosophical Magazine*, 5 (1852), 498–515, on 500.
29. Willem Hackmann, 'The Induction Coil in Medicine and Physics', in Christine Blondel, Françoise Parot, Anthony Turner and Mari Williams (eds), *Studies in the History of Scientific Instruments* (London: Roger Turner, 1989), pp. 235–50; Paolo

Brenni, 'Large Induction Coils', *Bulletin of the Scientific Instrument Society*, 125 (2015), 2–13.
30. Iwan Rhys Morus, 'Marketing the Machine: The Construction of Electrotherapeutics as Viable Medicine in Early Victorian England', *Medical History*, 36 (1992), 34–52.
31. Grove, 'On the Electro-chemical Polarity of Gases', 500–1.
32. Grove, 'On the Electro-chemical Polarity of Gases', 500.
33. John P. Gassiot, 'On Some Experiments made with Ruhmkorff's Induction Coil', *Philosophical Magazine*, 7 (1854), 97–9, on 99.
34. John P. Gassiot, 'On the Stratifications and Dark Bands in Electrical Discharges as Observed in Toricellian Vacua', *Philosophical Transactions*, 148 (1858), 1–16.
35. William Robert Grove, 'On the Striae seen in the Electric Discharge in Vacuo', *Philosophical Magazine*, 16 (1858), 18–22, on 20.
36. Grove, 'On the Striae seen in the Electric Discharge in Vacuo', 21.
37. Minute Book of the Philosophical Club, Royal Society Library MS.721, p. 17.
38. Minute Book of the Philosophical Club, Royal Society Library MS.721, p. 16.
39. Minute Book of the Philosophical Club, Royal Society Library MS.721, p. 17. See also Bonney, *Annals of the Philosophical Club of the Royal Society*, p. 27.
40. Bonney, *Annals of the Philosophical Club of the Royal Society*, p. 32.
41. Bonney, *Annals of the Philosophical Club of the Royal Society*, p. 36.
42. Minute Book of the Philosophical Club, Royal Society Library, MS.721, pp. 98–100.
43. Earl of Rosse, 'Presidential Address', *Proceedings of the Royal Society*, 6 (1854), 343–72, on 352.
44. Minute Book of the Philosophical Club, Royal Society Library, MS.721, pp. 98–100.
45. William Robert Grove to Lord Wrottesley, 29 January 1854, GCRI; *Report of the British Association for the Advancement of Science*, 25 (1855), p. xlvi.
46. S. H. Cradock to William Robert Grove, 31 March 1855, GCRI. Cradock was the principal of Brasenose College, Oxford, Grove's old college there. For Brodie's election to the chair of chemistry, see Robert Williams, Allan Chapman and John Rowlinson, *Chemistry at Oxford* (London: RSC Publishing, 2009), pp. 96–8.

Chapter 7

1. William Robert Grove to Lord Wrottesley, 29 January 1854, GCRI.
2. Charles Dickens, 'The Demeanour of Murderers', *Household Words*, 13 (1856), 505–7, on 505.
3. Stephen Bates, *The Poisoner* (London: Duckworth, 2015).
4. 'Photographic Society', *Photographic Journal*, 1 (1854), 2–5.
5. William Robert Grove, 'On a Voltaic Process for Etching Daguerreotype Plates', *Proceedings of the London Electrical Society*, 2 (1843), 94–100.
6. William Robert Grove, *On the Correlation of Physical Forces* (London: London Institution, 1846), p. 28.
7. William Henry Fox Talbot to William Robert Grove, December 1854, GCRI.
8. [William Robert Grove], 'Physical Science in England', *Blackwood's Magazine*, 54 (1843), 514–25, on p. 521.

9. [Grove], 'Physical Science in England', p. 521.
10. H. I. Dutton, *The Patent System and Inventive Activity during the Industrial Revolution, 1750–1852* (Manchester: Manchester University Press, 1984); Adrian Johns, *Piracy: The Intellectual Property Wars from Gutenberg to Gates* (Chicago: University of Chicago Press, 2009).
11. William Henry Fox Talbot to William Robert Grove, 17 December 1856, GCRI.
12. William Robert Grove, 'Suggestions for Improvements in the Administration of the Patent Law', *Jurist*, 6 (1860), 19–25, on 21.
13. *Report of the Royal Commissioners Appointed to Inquire into the Working of the Law Relating to Letters Patent for Inventions* (London: Her Majesty's Stationery Office, 1865), p. xii.
14. 'The Inaugural Address', *Nottinghamshire Guardian*, 24 August 1866, supplement.
15. 'The Inaugural Address'.
16. 'The Inaugural Address'.
17. William Robert Grove, 'Presidential Address', *Report of the British Association for the Advancement of Science*, 26 (1866), pp. liii–lxxxi, on pp. lv–lvi.
18. Grove, 'Presidential Address', p. lvi.
19. Grove, 'Presidential Address', p. lxiv.
20. Grove, 'Presidential Address', p. lxx.
21. Grove, 'Presidential Address', p. lxxviii.
22. 'British Association', *Nottinghamshire Guardian*, 24 August 1866, 5.
23. Editorial, *The Times*, 24 August 1866, 8.
24. 'The British Association', *Pall Mall Gazette*, 25 August 1866, 1–2, on 2.
25. Editorial, *Daily News*, 24 August 1866, 4.
26. Charles Darwin to Joseph Dalton Hooker, 30 August 1866. Darwin Correspondence Project, *darwinproject.ac.uk* (accessed 29 March 2016).
27. Joseph Dalton Hooker to Charles Darwin, 4 September 1866. Darwin Correspondence Project, *darwinproject.ac.uk* (accessed 29 March 2016).
28. William Robert Grove to Charles Darwin, 31 August 1866. Darwin Correspondence Project, *darwinproject.ac.uk* (accessed 29 March 2016).
29. Joseph Dalton Hooker to Charles Darwin, 29 May 1866. Darwin Correspondence Project, *darwinproject.ac.uk* (accessed 29 March 2016).
30. Joseph Dalton Hooker to Charles Darwin, 30 August 1866. Darwin Correspondence Project, *darwinproject.ac.uk* (accessed 29 March 2016).
31. Editorial, *Daily News*, 24 August 1866, 4.
32. William Robert Grove, *The Correlation of Physical Forces*, 5th edn (London: Longmans, Green and Co., 1867), p. viii.
33. William Robert Grove, *The Correlation of Physical Forces*, 6th edn (London: Longmans, Green and Co., 1874), p. vi.
34. William Robert Grove, 'An Address on the Importance of the Study of Physical Science in Medical Education', *British Medical Journal*, 57 (1869), 485–7, on 486.
35. William Robert Grove, 'Letter from the Honourable Mr. Justice Grove', *Report of the Royal Commission on Scientific Instruction and the Advancement of Science* (London: Her Majesty's Stationery Office, 1874), Appendix XVI, p. 71.

36. William Robert Grove, 'Antagonism', *Proceedings of the Royal Institution*, 12 (1887–9), 284–99, on 294.
37. 'Sir William Grove', *Daily News*, 4 August 1896.
38. 'Death of Sir William Grove', *The Times*, 3 August 1896, 8.
39. 'Sir William Grove', *Daily News*.
40. 'Death of Sir William Grove', *Cambrian*, 7 August 1896, 7.
41. 'The Late Sir William Grove: Proposed Bust at the Royal Institution', *Cambrian*, 14 August 1896, 4.
42. 'Sir William Grove Dead', *Evening Express*, 4 August 1896, 2.

Afterword

1. 'Death of Sir William Grove', *Cambrian*, 7 August 1896, 7.
2. 'Will of Sir William Robert Grove', *Evening Express*, 20 November 1896, 3.
3. Henry Minchin Noad, *Lectures on Electricity, Comprising Galvanism, Magnetism, Electro-magnetism, Magneto- and Thermo-electricity* (London: George Knight and Sons, 1844), frontispiece.
4. William Robert Grove, "New Voltaic Battery – Gaseous Elements", *Literary Gazette*, 26 (1842), 833.
5. Crosbie Smith, *The Science of Energy: A Cultural History of Energy Physics in Victorian Britain* (London: Athlone Press, 1998).
6. Peter Guthrie Tait and Balfour Stewart, *The Unseen Universe, or Physical Speculations on a Future State* (London: Macmillan, 1875), pp. 112–13.
7. Marie Boas Hall, *All Scientists Now* (Cambridge: Cambridge University Press, 1984); Donald Cardwell, *The Organisation of Science in England* (London: Heinemann, 1972).

BIBLIOGRAPHY

Primary sources

Ashman, J., 'An Artificial Leg, of an Improved Construction', *Report of the British Association for the Advancement of Science*, 18 (1848), 117.
Babbage, Charles, *Reflections on the Decline of Science in England* (London: B. Fellows, 1830).
——, *On the Economy of Machinery and Manufactures* (London: Charles Knight, 1832).
——, *The Ninth Bridgewater Treatise: A Fragment* (London: John Murray, 1837).
Barlow, Peter, *Treatise on the Manufactures and Machinery of Great Britain* (London: Baldwin & Cradock, 1831).
Bate, Spence, 'On Fossil Remains Recently Discovered in Bacon Hole, Gower; also Other Remains from Beneath the Bed of the River Tawey', *Report of the British Association for the Advancement of Science*, 18 (1848), 62–3.
Benson, Starling, 'On the Relative Positions of the Various Qualities of Coal in the South Wales Coal-measures', *Report of the British Association for the Advancement of Science*, 18 (1848), 65–6.
Brown, Thomas, *An Inquiry into the Relation of Cause and Effect* (Edinburgh: Archibald Constable, 1806).
Bulwer-Lytton, Edward, *The Coming Race* (1871; Stroud: Alan Sutton Publishing, 1995).
Butler, Charles, *The Inaugural Oration, Spoken on the 4th Day of November 1815, at the Ceremony of Laying the First Stone of the London Institution for the Diffusion of Science and Literature* (London: Longman, Hurst, Rees, Orme and Brown, 1816).
Carpenter, William Benjamin, 'On the Mutual Relations of the Vital and Physical Forces', *Philosophical Transactions*, 140 (1850), 727–57.
Chambers, Robert, *Vestiges of the Natural History of Creation and Other Evolutionary Writings*, ed. James Secord (1844; Chicago: University of Chicago Press, 1994).

Copleston, William James, *Memoir of Edward Copleston, D.D., Bishop of Llandaff* (London: John W. Parker & Son, 1851).

Daniell, John Frederic, 'On the Constant Voltaic Battery', *Philosophical Magazine*, 20 (1842), 294–304.

de la Beche, Henry, 'On the Geology of Portions of South Wales, Gloucestershire and Somersetshire', *Report of the British Association for the Advancement of Science*, 18 (1848), 79.

Dickens, Charles, 'The Demeanour of Murderers', *Household Words*, 13 (1856), 505–7.

——, *The Mudfog Papers* (London: Richard Bentley and Son, 1880).

Eliot, George, *Middlemarch: A Study in Provincial Life*, 8 vols (London: William Blackwood & Sons, 1872).

——, *The Mill on the Floss*, Great Writers edn (1860; London: Marshall Cavendish, 1986).

Elton, Professor, 'On the Ante-Columbian Discovery of America', *Report of the British Association for the Advancement of Science*, 1848, 18: 95-96.

Fletcher, Joseph, 'Statistics of Brittany and the Bretons', *Report of the British Association for the Advancement of Science*, 18 (1848), 114.

Galton, Francis, *English Men of Science: Their Nature and Nurture* (London: Macmillan, 1874).

Gassiot, John Peter, 'On Some Experiments made with Ruhmkorff's Induction Coil', *Philosophical Magazine*, 7 (1854), 97–9.

——, 'On the Stratifications and Dark Bands in Electrical Discharges as Observed in Toricellian Vacua', *Philosophical Transactions*, 148 (1858), 1–16.

Grove, William Robert, 'On a New Voltaic Combination', *Philosophical Magazine*, 13 (1838), 430–1.

——, 'On a New Voltaic Battery, and on Voltaic Combinations and Arrangements', *Philosophical Magazine*, 14 (1839), 287–93.

——, 'On Voltaic Series and the Combination of Gases by Platinum', *Philosophical Magazine*, 14 (1839), 127–30.

——, 'On Some Phenomena of the Voltaic Disruptive Discharge', *Philosophical Magazine*, 16 (1840), 478–82.

——, 'New Voltaic Battery – Gaseous Elements', *Literary Gazette*, 26 (1842), 833.

——, *On the Progress of Physical Science since the Opening of the London Institution* (London: London Institution, 1842).

——, 'Remarks on a Letter of Prof. DANIELL, in the Philosophical Magazine, "On the Constant Voltaic Battery"', *Philosophical Magazine*, 21 (1842), 333–5.

——, 'Experiments on Voltaic Reaction', *Philosophical Magazine*, 23 (1843), 45–7.

——, 'On the Subject of Mr. DANIELL's Last Communication', *Philosophical Magazine*, 22 (1843), 32–5.
——, 'Physical Science in England', *Blackwood's Magazine*, 54 (1843), 514–25.
——, 'On Photography', *Reports of the British Association for the Advancement of Science*, 14 (1844), 37.
——, 'On the Progress Made in the Application of Electricity as a Motive Power', *Literary Gazette*, 28 (1844), 113.
——, 'On the Application of Voltaic Ignition to Lighting Mines', *Philosophical Magazine*, 27 (1845), 442–6.
——, *On the Correlation of Physical Forces* (London: London Institution, 1846).
——, 'On Certain Phenomena of Voltaic Ignition and the Decomposition of Water into its Constituent Gases by Heat', *Philosophical Transactions*, 137 (1847), 1–16.
——, 'On the Peculiar Cooling Effects of Hydrogen and its Compounds in Cases of Voltaic Ignition', *Report of the British Association for the Advancement of Science*, 18 (1848), 55.
——, 'On the Effect of Surrounding Media on Voltaic Ignition', *Philosophical Transactions*, 139 (1849), 49–59.
——, 'On the Electro-chemical Polarity of Gases', *Philosophical Magazine*, 5 (1852), 498–515.
——, 'On the Striae seen in the Electric Discharge in Vacuo', *Philosophical Magazine*, 16 (1858), 18–22.
——, 'Suggestions for Improvements in the Administration of the Patent Law', *Jurist*, 6 (1860), 19–25.
——, 'Presidential Address', *Report of the British Association for the Advancement of Science*, 26 (1866), liii–lxxxi.
——, *The Correlation of Physical Forces*, 5th edn (London: Longmans, Green and Co., 1867).
——, 'An Address on the Importance of the Study of Physical Science in Medical Education', *British Medical Journal*, 57 (1869), 485–7.
——, 'Letter from the Honourable Mr. Justice Grove', *Report of the Royal Commission on Scientific Instruction and the Advancement of Science* (London: Her Majesty's Stationery Office, 1874).
——, *The Correlation of Physical Forces*, 6th edn (London: Longmans, Green and Co., 1874).
——, 'Antagonism', *Proceedings of the Royal Institution*, 12 (1887–9), 284–99.
Herschel, John, *Preliminary Discourse on the Study of Natural Philosophy* (London: Longman, Rees, Orme, Brown & Green, 1831).
Hill, Benjamin, 'On a New Electro-magnetic Machine', *Proceedings of the London Electrical Society*, 2 (1841), 83–6.

Historical Account of the London Institution (London: London Institution, 1835).

Hunt, Robert, 'On Electro-magnetism as a Motive Power', *Proceedings of the Institution of Civil Engineers*, 16 (1857), 386–421.

Joule, James Prescott, 'On the Mechanical Equivalent of Heat, and the Constitution of Elastic Fluids', *Report of the British Association for the Advancement of Science*, 18 (1848), 21–2.

Kingsley, Charles, *The Water Babies* (London: Macmillan, 1863).

Lankester, Edwin, 'On Some Vegetable Monstrosities Illustrating the Laws of Morphology', *Report of the British Association for the Advancement of Science*, 18 (1848), 85–6.

Laycock, Thomas, *Mind and Brain, or, the Correlations of Consciousness and Organisation* (London: Simpkin, Marshall & Co., 1859).

Lyell, Katherine, *Life, Letters and Journals of Sir Charles Lyell*, 2 vols (London: John Murray, 1881).

——, *Memoir of Leonard Horner*, 2 vols (London: Women's Printing Society, 1890).

Mackintosh, Thomas Simmons, *The Electrical Theory of the Universe* (Boston: Josiah P. Mendum, 1846).

Matteucci, Carlo, 'On the Electricity of Flame, with Comments by Mr. Grove', *Philosophical Magazine*, 8 (1854), 399–404.

Maxwell, James Clerk, 'Grove's Correlation of Physical Forces', *Nature*, 10 (1874), 302–4.

Moggridge, Matthew, 'On Two Cases of Uncommon Atmospheric Refraction', *Report of the British Association for the Advancement of Science*, 18 (1848), 33–4.

——, 'On a Peculiarity in the Protococcus Nivalis', *Report of the British Association for the Advancement of Science*, 18 (1848), 86.

Nasmyth, James, 'On a Peculiar Property of Coke', *Report of the British Association for the Advancement of Science*, 18 (1848), 56.

Noad, Henry Minchin, *Lectures on Electricity, Comprising Galvanism, Magnetism, Electro-magnetism, Magneto- and Thermo-electricity* (London: George Knight and Sons, 1844).

Northampton, marquis of, 'Presidential Address', *Proceedings of the Royal Society*, 5 (1843–50), 698–703.

——, 'Presidential Address', *Proceedings of the Royal Society*, 5 (1843–50), 761–7.

——, 'Presidential Address', *Report of the British Association for the Advancement of Science*, 18 (1848), xxxi–xxxix.

Owen, Richard, 'On the Communication between the Tympanum and the Palate in the Crocodiles', *Report of the British Association for the Advancement of Science*, 18 (1848), 79–80.

William Robert Grove

Portraits of Men of Eminence in Literature, Science, and Art, with Biographical Memoirs, the Photographs from Life, by Ernest Edwards, BA (London: Alfred William Bennett, 1865).

Report of the Royal Commissioners Appointed to Inquire into the Working of the Law Relating to Letters Patent for Inventions (London: Her Majesty's Stationery Office, 1865).

Rosse, earl of, 'Presidential Address', *Proceedings of the Royal Society*, 6 (1854), 343–72.

Smee, Alfred, *Elements of Electro-metallurgy, or the Art of Working in Metals by the Galvanic Fluid* (London: Longman, Rees, Orme, Brown & Longman, 1841).

Smith, Albert, *Gavarni in London: Sketches of Life and Character* (London: David Bogue, 1849).

Somerville, Mary, *On the Connexion of the Physical Sciences*, 2nd edn (London: John Murray, 1835).

Sturgeon, William, 'Account of Improved Electro-magnetic Apparatus', *Annals of Philosophy*, 12 (1826), 357–61.

Ure, Andrew, *The Philosophy of Manufactures* (London: Charles Knight, 1835).

Walker, Charles Vincent, 'Notice on the Relative Powers of certain Diaphragm Voltaic Combinations, and on a New Form of the Negative Element in Voltaic Arrangements', *Proceedings of the London Electrical Society*, 2 (1841), 114–18.

Ward, W. S., 'On a Galvanometer', *Report of the British Association for the Advancement of Science*, 18 (1848), 62.

Whewell, William, *History of the Inductive Sciences*, 3 vols (London: John Parker, 1837).

Williams, Cadogan, 'On the Desirability of Extending to the Working Classes the opportunity of Purchasing Deferred Annuities, as a Provision for Old Age', *Report of the British Association for the Advancement of Science*, 18 (1848), 105.

Williams, Thomas, 'On the Physical Conditions Regulating the Vertical Distribution of Animals in the Atmosphere and the Sea', *Report of the British Association for the Advancement of Science*, 18 (1848), 83.

Wishaw, Francis, 'On the Subaqueous Rope for Telegraphic and other Purposes', *Report of the British Association for the Advancement of Science*, 18 (1848), 123.

Yates, Edmund, *Recollections and Experiences*, 2 vols (London: Richard Bentley & Son, 1884).

Secondary sources

100 Welsh Heroes (Aberystwyth: Culturenet Cymru, 2004).

Altick, Richard, *The Shows of London* (Cambridge, MA: Belknap Press, 1978).

Barrie, D. S. M., *Regional History of the Railways of Great Britain: South Wales* (London: David and Charles, 1980).

Bates, Stephen, *The Poisoner* (London: Duckworth, 2015).

Beanland, W. A., *The History of the Royal Institution of South Wales* (Swansea: Royal Institution of South Wales, 1935).

Becher, Harvey, 'Radicals, Whigs and Conservatives: The Middle and Lower Classes in the Analytical Revolution in Cambridge in the Age of Aristocracy', *British Journal for the History of Science*, 28 (1995), 405–26.

Berman, Morris, *Social Change and Scientific Organization: The Royal Institution, 1799–1844* (London: Heinemann, 1978).

Boas Hall, Marie, *All Scientists Now* (Cambridge: Cambridge University Press, 1984).

——, *Promoting Experimental Learning: Experiment and the Royal Society, 1660–1727* (Cambridge: Cambridge University Press, 1991).

Bonney, Thomas, *Annals of the Philosophical Club of the Royal Society* (London: Macmillan, 1919).

Brenni, Paolo, 'Large Induction Coils', *Bulletin of the Scientific Instrument Society*, 125 (2015), 2–13.

Cardwell, Donald, *The Organisation of Science in England* (London: Heinemann, 1972).

Carlyle, Alexander (ed.), *New Letters of Thomas Carlyle* (London: Bodley Head, 1904).

Cooper, M. L., and V. M. D. Hall, 'William Robert Grove and the London Institution, 1841–1845', *Annals of Science*, 39 (1982), 229–54.

Corsi, Pietro, *Science and Religion: Baden Powell and the Anglican Debate, 1800–1860* (Cambridge: Cambridge University Press, 1988).

Crowther, J. G., *Statesmen of Science* (London: Cresset Press, 1965).

Desmond, Adrian, *The Politics of Evolution* (Chicago: University of Chicago Press, 1989).

Dutton, H. I., *The Patent System and Inventive Activity during the Industrial Revolution, 1750–1852* (Manchester: Manchester University Press, 1984).

Fox, Robert and Graeme Gooday (eds), *Physics in Oxford, 1839–1939* (Oxford: Oxford University Press, 2005).

Gascoigne, John, *Joseph Banks and the English Enlightenment* (Cambridge: Cambridge University Press, 1994).

Hackmann, Willem, 'The Induction Coil in Medicine and Physics', in Christine Blondel, Françoise Parot, Anthony Turner and Mari Williams (eds), *Studies in the History of Scientific Instruments* (London: Roger Turner, 1989), pp. 235–50.

Hays, J. N., 'Science in the City: The London Institution, 1819–1840', *British Journal of the History of Science*, 7 (1974), 46–62.

——, 'The London Lecturing Empire, 1800–50', in Ian Inkster and Jack Morrell (eds), *Metropolis and Province: Science in British Culture 1780–1850* (London: Hutchison, 1983), pp. 91–119.

Hoogers, Gregor, *Fuel Cell Technology Handbook* (Boca Raton: CRC Press, 2002).

Hunt, Bruce, *The Maxwellians* (Ithaca: Cornell University Press, 1994).

Johns, Adrian, *Piracy: The Intellectual Property Wars from Gutenberg to Gates* (Chicago: University of Chicago Press, 2009).

Lyons, Henry, *Record of the Royal Society* (London: Royal Society, 1940).

——, *The Royal Society 1660–1940: A History of its Administration under its Charters* (Cambridge: Cambridge University Press, 1944).

Miller, David P., 'Between Hostile Camps: Sir Humphry Davy's Presidency of the Royal Society of London, 1824–27', *British Journal for the History of Science*, 16 (1983), 1–47.

——, 'Method and the Micropolitics of Science: The Early Years of the Geological and Astronomical Societies', in John Schuster and Richard Yeo (eds), *The Politics and Rhetoric of Scientific Method* (Dordrecht: D. Reidel & Co., 1986), pp. 227–58.

Miskell, Louise, *Intelligent Town: An Urban History of Swansea, 1780–1855* (Cardiff: University of Wales Press, 2006).

Mollan, Charles (ed.), *William Parsons, 3rd Earl of Rosse* (Manchester: Manchester University Press, 2016).

Morrell, Jack, and Arnold Thackray, *Gentlemen of Science: Early Years of the British Association for the Advancement of Science* (Oxford: Oxford University Press, 1981).

Morus, Iwan Rhys, 'The Sociology of Sparks: An Episode in the History and Meaning of Electricity', *Social Studies of Science*, 18 (1988), 387–417.

——, 'Different Experimental Lives: Michael Faraday and William Sturgeon', *History of Science*, 30 (1992), 1–28.

——, 'Marketing the Machine: The Construction of Electrotherapeutics as Viable Medicine in Early Victorian England', *Medical History*, 36 (1992), 34–52.

——, 'Currents from the Underworld: Electricity and the Technology of Display in early Victorian London', *Isis*, 84 (1993), 50–69.

——, 'Worlds of Wonder: Sensation and the Victorian Scientific Performance', *Isis*, 101 (2010), 806–16.

——, *Frankenstein's Children: Electricity, Exhibition and Experiment in Early Nineteenth-century London* (Chicago: University of Chicago Press, 1998).

——, *When Physics Became King* (Chicago: University of Chicago Press, 2005).

——, Simon Schaffer and James Secord, "Scientific London", in Celina Fox (ed.), *London: World City, 1800–1840* (London and New Haven: Yale University Press, 1992), pp. 129–42.

Olson, Richard, *Scottish Philosophy and British Physics, 1750–1880* (Princeton: Princeton University Press, 1975).

Orange, A. D., *Philosophers and Provincials: The York Philosophical Society 1822–1844* (York: York Philosophical Society, 1973).

Peterson, M. Jeanne, *The Medical Profession in mid-Victorian London* (Berkeley and Los Angeles: University of California Press, 1978).

Rees, Ronald, *King Copper: South Wales and the Copper Trade, 1584–1895* (Cardiff: University of Wales Press, 2000).

Ross, Sydney, 'Scientist: The Story of a Word', *Annals of Science*, 19 (1962), 65–85.

Secord, James, *Victorian Sensation: The Extraordinary Publication, Reception and Secret Authorship of Vestiges of the Natural History of Creation* (Chicago: University of Chicago Press, 2003).

——, *Visions of Science: Books and Readers at the Dawn of the Victorian Age* (Oxford: Oxford University Press, 2014).

Silver, Harold, *English Education and the Radicals 1780–1850* (London: Routledge and Kegan Paul, 1977).

Smith, Crosbie, *The Science of Energy: A Cultural History of Energy Physics in Victorian Britain* (London: Athlone Press, 1998).

Stewart, Larry, *The Rise of Public Science* (Cambridge: Cambridge University Press, 1992).

Toman, John, *Kilvert's World of Wonders: Growing Up in Mid-Victorian England* (Cambridge: Lutterworth Press, 2013).

Wilks, Samuel, and George Bettany, *A Biographical Dictionary of Guy's Hospital* (London: Ward, Lock, Bowden and Co., 1892).

Williams, L. Pearce (ed.), *The Selected Correspondence of Michael Faraday*, 2 vols (Cambridge: Cambridge University Press, 1971).

Williams, Robert, Allan Chapman and John Rowlinson, *Chemistry at Oxford* (London: RSC Publishing, 2009).

INDEX

A
Accum, Frederick 25
Adelaide Gallery 26–7, 44–5
Airy, George Bidell 15
Armstrong, William George 44
Astronomical Society 67

B
BAAS (British Association for the Advancement of Science) 19, 21–2, 23, 31, 40, 85–101, 127–33, 145
Babbage, Charles 36, 50, 51, 54, 58, 67, 70, 86, 104
Bacon, Francis 52, 64
Bain, Alexander 62
Banks, Joseph 66–7
Barlow, Peter 59
Bate, Spence 96
battery, nitric acid 3, 6, 23, 31–2, 42, 43, 56, 60, 87, 100–1, 116
battery, voltaic 3, 6, 20–1, 23, 27, 29–31, 37–9, 42, 43, 46, 56, 60, 87, 100–1, 116, 134
Baxter, Thomas 7
Beche, Henry de la 18, 40, 83, 91, 96, 105, 107–8
Bell, Thomas 109–11
Benson, Starling 96
Bevan, Anne 7
Birkbeck, George 24
Bowman, William 110–11
Brasenose College 10, 11
Brewster, David 91, 95
Brown, Thomas 57
Buckland, William 19
Bulwer-Lytton, Edward 63
Burlington House 119
Butler, Charles 34–5

C
Callan, Nicholas 114
Cardiff 11, 90
Carpenter, William Benjamin 62, 111
Chemical Society 119
Christie, Samuel Hunter 108
Clarke, Edward Marmaduke 27
Coleridge, Samuel Taylor 86
Coliseum 27
Conybeare, William 15
Cooke, William Fothergill 44, 125

D
Daniell, John Frederic 31, 42, 71–2

Darwin, Charles 129–33
Davenport, Thomas 45
Davy, Edward 44–5
Davy, Humphry 24, 67–8
Dickens, Charles 63, 87
Dillwyn, Lewis Weston 7, 10, 13–14, 17–18, 19, 22, 28, 89, 91–2, 96, 101
discharge tube experiments 112–16

E

Electrical Society of London 26–7, 30, 34, 40, 100
electricity 20, 26–7, 29–30, 36, 38–9, 43–7, 95, 112–16
electromagnet 27
electromagnetic engine 27, 36, 43–5, 100
electrometallurgy 36, 43, 45–6
Eliot, George 24, 62
Engels, Friedrich 63

F

Faraday, Michael 24, 26–7, 30, 32, 42, 63, 71–2, 107, 112, 114
Fisher, George Thomas 33
Forbes, Edward 105–7
fuel cell *see* gas battery

G

galvanism 20, 31
gas battery 1–3, 4, 39–40, 55–6, 141
Gassiot, John Peter 33–4, 40, 43–4, 46, 105, 112, 115–16
Gassiot's cascade 115
Geological Society 67, 74
Gilbert, Davies 68

Gower 7, 9, 96–7, 100
Graham, Thomas 21, 33, 87, 105
Griffiths, Evan 9–10
Grove, Coleridge 32, 139
Grove, Florence Craufurd 29, 139
Grove, John 7–8, 10–11
Grove, Thomas 8
Grove, William 8–9, 10, 12, 17
Grove, William Robert (senior) 8
Grove, William Robert
  antagonism, lecture on 134
  BAAS, presidency of 127–33
  BAAS, Swansea meeting 87–91, 92–8, 100–1
  birth 7
  continuity, lecture on 128–33
  death 136–7
  discharge phenomena, experiments on 112–26
  education 9–11, 18–19, 23
  gas battery, invention of 38–9
  London Institution, chair 33, 35–42
  marriage 28–9
  natural selection, support for 129–33
  nitric acid battery, invention of 21, 31–2
  On the Correlation of Physical Forces 52–63, 133–4, 143
  patents, views on 123–6
  and the Philosophical Club 104–12, 116–20
  progress of the physical sciences, lecture on 35–7, 53–4
  Royal Institution of South Wales 14, 18–22
  Royal Society, fellow of 32
  Royal Society, reform of 71–81
  wealth 137, 139

H

Herschel, John 4, 48–9, 51, 52, 67, 70, 105, 107
Hill, Benjamin 10, 100
Hooker, Joseph Dalton 131–3
Horner, Leonard 74–81, 91, 105, 109, 111
Hume, David 57

I

induction coil 113–16
industrial revolution 1, 13–14

J

Jacobi, Hermann von 32, 45
Jeffreys, John Gwyn 96
Jerdan, William 39
Joule, James Prescott 27, 95, 141
juxtaposition 116–20

K

Kilvert, Revd J. 10
Kingsley, Charles 4, 63

L

Lardner, Dionysius 48
Laroche, Martin 123
Laurels, The 7, 22, 29
Laycock, Thomas 62
Leithead, William 27
Linnean Society 67
Literary & Philosophical Institutes 14–16
Llewellyn, John Dillwyn 18, 22, 40, 89, 100

London Institution 25, 27, 33–42, 47, 52–3, 58, 60, 63, 112, 141, 144
London Mechanics Institute 24
Lyell, Charles 104–5, 109, 111–12

M

Mackintosh, Thomas Simmons 51–2
magic lantern 1, 26, 37
magneto-electric machine 26
Marx, Karl 63
Maxwell, James Clerk 63, 141, 143
Merthyr Literary & Scientific Institution 15–16
Merthyr Tydfil 15–17
Moggridge, Matthew 90, 95, 96
Mond, Ludwig 2
Murchison, Roderick Impey 104–5

N

natural selection 129–33
Neath Museum & Society for Promoting the Arts & Sciences 15, 16
Newton, Isaac 65
Noad, Henry Minchin 140–1
Northampton, marquis of 75, 81–4, 92–4, 99, 104, 106, 111

O

Owen, Richard 97, 104, 118
Oxford 10–11, 18–19, 86, 89, 120, 121

P

Palmer, William 122
Parliamentary Committee, BAAS 119–20

patent legislation 123–6
Pepys, William Haseldine 37
Phillips, John 88–90
Philosophical Club 104–12, 116–20
Photography 40, 123–6
Playfair, Lyon 40, 118
Powell, Baden 19
Powles, Emma Maria 28–9, 136
Powles, John Diston 28–9, 34
Priestley, Joseph 48, 66

R
Rayleigh, lord 2
Rebecca riots 17, 89
Reid, Thomas 57
Rive, Auguste de la 43
Roget, Peter Mark 75, 78–80, 104, 106, 108–9, 111
Rosse, earl of 107, 112, 118
Royal Institution 23–5, 27–8, 32, 34, 60, 112
Royal Institution of South Wales 14, 16–18, 20–1, 28–9, 31, 87–8, 89–90, 92, 95, 137, 144
Royal Polytechnic Institution 26–7, 44
Royal Society 6, 18, 32–3, 40–2, 47, 50, 62, 65–84, 104–12, 116–20, 126, 143–4
Ruhmkorff, Heinrich 113–15

S
Sabine, Edward 75, 78–9, 91, 118
Saxton, Joseph 26
Sloane, Hans 65–6
Somerville, Mary 49
steampunk 1
Sturgeon, William 25
Sussex, duke of 69–70

Swansea 6, 7–9, 11–18, 22, 23, 29, 61, 85–101, 112, 136–7, 144–5
Swansea Scientific & Literary Institution *see* Royal Institution of South Wales

T
Tait, Peter Guthrie 4, 63, 143
Talbot patent case 123–6
Talbot, William Henry Fox 18, 40, 123–6
Telegraph 3, 36, 44, 46, 98, 125, 141
Thomson, William 4, 141, 143
Tooke, William 83–4
Tyndall, John 4

U
Ure, Andrew 58–9

V
Vernon Harcourt, William 85–7
Vestiges of the Natural History of Creation 51, 53, 129
Vivian, John Henry 13–14, 18, 19, 28, 33, 92

W
Wakley, Thomas 70
Wales 3, 5–6, 88, 90–1, 145
Walker, Charles Vincent 38, 46
Welsh science 3, 90–1, 144–5
Wheatstone, Charles 33, 44, 125
Whewell, William 49–50, 51, 52, 86, 95
Williams, Thomas 97
Wrottesley, lord 119–20